Fundamentals of Electromagnetics 2: Quasistatics and Waves

Fundamentals of Electromagnetics 2: Quasistatics and Waves
David Voltmer

ISBN: 978-3-031-00572-5 paperback
ISBN: 978-3-031-00572-5 paperback

ISBN: 978-3-031-01700-1 ebook
ISBN: 978-3-031-01700-1 ebook

DOI 10.1007/978-3-031-01700-1

A Publication in the Springer series
SYNTHESIS LECTURES ON COMPUTATIONAL ELECTROMAGNETICS #15

Lecture #15
Series Editor: Constantine A. Balanis, Arizona State University

Library of Congress Cataloging-in-Publication Data

Series ISSN: 1932-1252   print
Series ISSN: 1932-1716   electronic

First Edition
10 9 8 7 6 5 4 3 2 1

Fundamentals of Electromagnetics 2: Quasistatics and Waves

David Voltmer
Rose-Hulman Institute of Technology

SYNTHESIS LECTURES ON COMPUTATIONAL ELECTROMAGNETICS #15

ABSTRACT

This book is the second of two volumes which have been created to provide an understanding of the basic principles and applications of electromagnetic fields for electrical engineering students. *Fundamentals of Electromagnetics Vol 2: Quasistatics and Waves* examines how the low-frequency models of lumped elements are modified to include parasitic elements. For even higher frequencies, wave behavior in space and on transmission lines is explained. Finally, the textbook concludes with details of transmission line properties and applications. Upon completion of this book and its companion *Fundamentals of Electromagnetics Vol 1: Internal Behavior of Lumped Elements*, with a focus on the DC and low-frequency behavior of electromagnetic fields within lumped elements, students will have gained the necessary knowledge to progress to advanced studies of electromagnetics.

KEYWORDS

Electromagnetics Introduction, Electromagnetic Fields, Quasistatics, Waves, Wave behavior in space and on transmission lines, transmission line properties and applications, low-frequency models of lumped elements

Contents

Preface

"Learning about 0's and 1's is much easier than learning about Electromagnetics."
—Robert MacIntosh

Most students begin their studies in electrical engineering with courses in computer logic and electric circuits. This is quite understandable since the students' mathematical background is usually limited to scalar mathematics. But, the fundamental principles of electrical engineering upon which the majority of physical processes such as semiconductor devices, power generation, or wireless technology are represented mathematically by three-dimensional, vector calculus. Electromagnetics is a very challenging subject for students lacking a strong background in mathematics or three-dimensional visualization skills—even for students who have a thorough understanding of electric circuits. On the other hand, the principles of electromagetics are relatively simple when the underlying vector calculus is understood by the students. This textbook is a non-traditional approach to bridge this gap that is based upon the similarities of lumped passive elements of circuits—resistors, capacitors, and inductors.

The terminal characteristics of lumped elements are derived from the observable quantities of voltage difference and current flow that are measured with instruments of finite size, i.e., dimensions on a macro-scale. But a more fundamental description of the underlying phenomena is on a micro-scale, infinitesimal in size, i.e., smaller than the smallest instrument. This behavior is described in terms of differential dimensions, with dimensions that shrink to zero; it is not measurable by man-made instruments and is much more complicated than two terminal properties of voltage drop and current flow. These inner workings of lumped elements on a micro-scale are related to their external, observable macro-scale (or terminal) characteristics.

This approach utilizes the following set of underlying restrictions:

- The terminal behavior of passive, lumped elements, $v_R = i_R R$, $i_C = C dv_C/dt$, and $v_L = L di_L/dt$, is known from circuits. Circuit theory experience, especially with KVL and KCL, is assumed and used.

- Each material type has only a single electromagnetic property—conductors have non-zero conductivity, σ, and allow current flow; dielectrics have permittivity, ε, and store only electric energy; and magnetic materials have permeability, μ, and store only magnetic energy.

- Lumped elements of finite size can be decomposed into discrete, curvilinear regions each of which behaves as an incremental lumped element. The total element value is calculated using the series/parallel equations of circuits.

- The electric or magnetic scalar potentials of all three types of lumped elements satisfies Laplace's equation, $\nabla^2 V = 0$. The same solution methods can be used for all elements.

- There are no fringing or leakage fields from any lumped elements. Consequently, all inductors have torroidal-shaped, magnetic cores.

- The frequency of operation is low enough that the elements are small compared to wavelength. Consequently, radiation and wave effects are ignored.

- Linearity and superposition apply with one notable exception—magnetic circuits.

- All electrodes and connecting wires are lossless with zero resistance and as such are called Perfect Electric Conductors (PEC).

These restrictions are imposed so that the basic principles of electromagnetics are not hidden by complicating geometric or mathematical details. Those details can come in following courses. Those students who take more advanced electromagnetic courses are not hindered by this approach. In fact, a case may be made that this approach provides a more complete understanding of the basics than traditional approaches.

Analytic expressions are used throughout this text as a calculation technique. Numeric and graphical methods for two dimensional structures are integrated within the textbook as well. In addition, circuit simulation is introduced as a viable solution method.

Finally, It is assumed that all students have studied electricity and magnetism in introductory physics and have an understanding of the concepts of force, displacement, work, potential energy, opposite charges attract, like charges repel, and current is usually assumed to be uniformly distributed currents in resistors. In addition, students should be familiar with vector algebra. Most work uses Cartesian coordinates with a few examples in cylindrical and spherical coordinates. The SI system of units is used throughout.

Dr. David R. Voltmer   May 2007
Professional Engineer
Professor Emeritus
Electrical and Computer Engineering
Rose-Hulman Institute of Technology

CHAPTER 5

Quasi-Static Fields

We have covered quite a bit of territory in the first four chapters of this text, but it is only the beginning. The restriction of those chapters that objects be small compared to wavelength is quite limiting; as we deal with higher frequencies (shorter wavelengths), our earlier models are no longer valid. Just what happens when the region in which we must solve for the fields is no longer a small fraction of wavelength? In a related way, how do fields behave in a region where they are not confined by PECs or by flux guides? This chapter will provide some introductory insight into these questions.

Circuit concepts provide an intuitive framework on which we build the more abstract structure of fields. However, the distributed nature of electromagnetic fields becomes more important as the regions of solution become larger. The vector mathematics becomes more important; terminal voltages and currents are not as helpful when there are really no elements to measure. So, boldly, we go where few have gone before.

5.1 FARADAY'S LAW REVISITED

The voltage drop across the open-circuited terminals of a PEC loop of wire is expressed in terms of the time-varying magnetic flux that is encircled by the loop as expressed by Faraday's law

$$V_{\text{LOOP}} = -\frac{d\Psi_m}{dt}. \qquad (5.1)$$

We learned in Chapter 3 that the polarity of the induced voltage is predicted by Lenz's law. The polarity of the induced voltage tends to establish a current that opposes the initial change in flux, a sort of electromagnetic inertia. These two principles provide an adequate basis for the study of inductors and magnetic devices. But, they can be extended to describe much more general electromagnetic behavior.

Since electric field intensity is the basis for a voltage difference in resistors and capacitors, it seems plausible that it is in inductors as well. Postulating the existence of an electric field intensity, we can express the voltage difference of the LHS of Eq. (5.1) in terms of electric field intensity as

$$V_{\text{LOOP}} = V_{\text{INIT}} - V_{\text{FINAL}} = \oint_{\text{CCW}} \mathbf{E} \cdot \mathbf{dl}. \tag{5.2}$$

The CCW direction is chosen to define a positive V_{LOOP} with the polarity shown in Fig. 5.1.

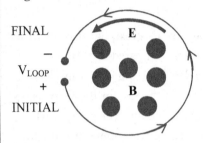

FIGURE 5.1: Induced voltage in a PEC wire loop and the accompanying **E** field.

In contrast to the conservative electric fields within resistors and capacitors, the electric field within an inductor is not conservative! A nonzero voltage drop means that the line integral does not vanish; that is why there is an induced voltage. Moreover, since the integral depends upon the path, i.e., the configuration of the wire, so does the induced voltage. This is to be expected since the flux enclosed by the wire depends upon its configuration.

I must admit that I "lied" in the earlier chapters when I said that electric fields are always conservative. They are so nearly conservative within resistors and capacitors that this assumption introduces negligible error. For inductors and magnetic circuits, we acknowledged that they were nonconservative, but, because we assumed that the frequency of operation was low, we didn't need to alter our approach much to produce accurate calculations. Now that the truth is out, we must get to the root of this phenomenon and determine the features of electric fields which give them such different characteristics in different situations.

The total voltage can be interpreted as the summation of all of the infinitesimal voltage drops $\Delta V = -\mathbf{E} \cdot \mathbf{\Delta l}$ around the PEC loop. There can be no voltage drop along a PEC wire even though the total voltage induced in the loop by the changing flux depends upon the configuration of the wire. This leads to the fact that the induced voltage is all observed across the terminals of the PEC wire. What if the wire had not been there? Then the integral along the path would have still given the same result with incremental induced voltages all along the path. However, there would be no measurable voltage between two closely spaced points, just a theoretical, incremental ΔV.

This brings up the question of how to measure a voltage drop without the position of the wires affecting the results. Only if there is insignificant magnetic flux enclosed by the wire leads that connect the voltmeter is the voltage independent of the wire position. This is the case of DC and slowly varying fields. When there is insignificant magnetic flux enclosed by the meter and its leads, then the electric field is essentially conservative. Consequently, any position of the wires gives the same result for the voltage drop between two points. On the other hand, that portion of the voltage that is due to a changing magnetic field depends upon the shape and position of the wire leads. The electric field that is associated with this *type* of voltage drop is not conservative.

The time rate of change of the enclosed magnetic flux, the RHS of Eq. (5.1), can be rewritten in terms of the magnetic flux density so that Faraday's law becomes

$$V_{\text{LOOP}} = \oint_{\text{CCW}} \mathbf{E} \cdot \mathbf{dl} = -\frac{d}{dt} \left(\iint_S \mathbf{B} \cdot \mathbf{ds} \right) \tag{5.3}$$

where the usual convention of the RH rule relates the line and surface integrals. The line integral of the electric field intensity is equal to the surface integral of the magnetic flux density that it encloses. This relationship is valid throughout all space for every path and all associated surfaces.

From Eq. (5.3), we can see that if the magnetic flux enclosed by the path L is unchanging, then the line integral will vanish, the electric field is conservative, and there is no induced voltage. This is the case for the electric fields of resistors and capacitors as described in Chapters 1 and 2 which are due predominantly to current and electric fluxes rather than magnetic flux. As shown on the RHS of Eq. (5.3), if the time rate of change of enclosed magnetic flux is small enough, the induced voltage around a closed loop can be considered zero. An interpretation of the RHS is simplest when the magnetic field has no spatial variation, is sinusoidal in time, and oriented perpendicular to the loop as in a magnetic core so that the total flux is expressed as $\Psi_m = AB_o \sin \omega t$. The time derivative of flux is expressed as $d\Psi_m/dt = d(AB)/dt = AdB/dt = \omega AB_o \cos \omega t$ when the loop is stationary. The magnitude of the magnetic flux density is important in determining the induced voltage. But the size of the loop and the frequency of the magnetic field are equally important. Making use of the well-known frequency–wavelength relationship of $f = c/\lambda$, we can rewrite the induced voltage as $d\Psi_m/dt = (2\pi c/\lambda)AB_o \cos \omega t = (A/\lambda)2\pi c B_o \cos \omega t$. The size of the loop relative to the wavelength of the magnetic field is an important parameter. For low frequencies, where the loop is relatively small compared to wavelength, the induced voltage can be neglected. This is typical of most circuits that operate below several MHz. But for higher frequencies, where the dimensions of the loop are on the same order of magnitude or smaller than the wavelength of the magnetic field, the induced voltage for the same sized circuit can be significant. When asked the question, "are electric fields conservative?", the answer is yes and no. Yes, if the dimensions of the circuits and frequency of the fields are small enough; no, if they are larger.

Assuming that $\mathbf{E}$ is sufficiently well behaved so that $\nabla \times E$ exists, we can apply Stokes' theorem to the LHS of Eq. (5.3) to give

$$\oint_{\text{CCW}} \mathbf{E} \cdot \mathbf{dl} = \iint_S \nabla \times \mathbf{E} \cdot \mathbf{ds} = -\frac{d}{dt} \left(\iint_S \mathbf{B} \cdot \mathbf{ds} \right). \tag{5.4}$$

For this portion of the course, assume that observations are made with a fixed loop, no change in size, orientation, or position. This means that the surface S is not changing with time so that the time derivative can be moved inside the integral, operating on $\mathbf{B}$ only. The time derivative is also modified to a partial derivative since $\mathbf{B}$ may have time-independent spatial variation as well. In addition, though an infinite number of surfaces can be chosen for both surface integrals of Eq. (5.4), the work is simplified by choosing the same surface for both. These two considerations enable us to rewrite Eq. (5.4) as

$$\iint_S \left(\nabla \times \mathbf{E} + \frac{\partial \mathbf{B}}{\partial t} \right) \cdot \mathbf{ds} = 0. \tag{5.5}$$

By shrinking the surface to infinitesimal dimensions, the integral can be approximated as

$$\iint_S \left(\nabla \times \mathbf{E} + \frac{\partial \mathbf{B}}{\partial t} \right) \cdot \mathbf{ds} \approx \left(\nabla \times \mathbf{E} + \frac{\partial \mathbf{B}}{\partial t} \right) \cdot \Delta \mathbf{s} = 0. \tag{5.6}$$

In the limit as the surface becomes small, $\Delta \mathbf{s} = \mathbf{a}_N |\Delta a|$ and $|\Delta \mathbf{s}| > 0$, so that the bracketed quantity must vanish, i.e.,

$$\nabla \times \mathbf{E} = -\frac{\partial \mathbf{B}}{\partial t}, \tag{5.7}$$

the differential form of Faraday's law. As expected, when the magnetic flux density at a point is unchanging, the electric field intensity is conservative and $\nabla \times \mathbf{E} = 0$. But, if the magnetic field intensity is time varying, then the electric field is not conservative. Since the curl of a vector is perpendicular to the vector itself, the electric field is perpendicular to the magnetic field at each point as implied by the integral form also.

The physically observable form of Faraday's law, Eq. (5.1), has been written in terms of electric and magnetic fields. Treating these fields as mathematical functions, they have been recast in differential form, Eq. (5.7). Though the latter form is not observable, it is very useful in predicting the behavior of electromagnetic fields. We will return to this form later.

Example 5.1-1. A time-varying current flows in a long, filamentary wire that is near a closed, PEC loop. Calculate the induced voltage in the loop. For calculation purposes, the rectangular loop has dimensions of 10 cm parallel and 5 cm perpendicular to the current and its nearest side is located 5 cm from the current. The current is 10 A operating at 60 Hz. The magnetic field established by the current is perpendicular to the plane of the loop and, from Ampere's law, it is expressed as $H = 10/2\pi\rho$. The total magnetic flux that links the PEC loop is given by $\Psi_m = [4\pi \times 10^{-7}(10)(0.1)/2\pi]\ln(0.1/0.05)(\sin 377t) = 0.139 \sin 377t$ μwb. The induced voltage is $V_{\text{LOOP}} = -52.3\mu$V; the smaller dimension of the loop is $10^{-8}\lambda$.

Example 5.1-2. Calculate the induced voltage for Example 5.1-1 if the source has a frequency of 1 MHz? The magnetic flux linking the PEC loop is given by $\Psi_m = [4\pi \times 10^{-7}(10)(0.1)/2\pi]\ln(0.1/0.05)(\sin 6.28 \times 10^6 t) = 0.139 \sin 6.28 \times 10^6 t$ V. The induced voltage is -0.87V, considerable larger than for 60 Hz; the smaller dimension of the loop is 0.00016λ.

Example 5.1-3. Calculate the induced voltage for Example 5.1-1 if the source has a frequency of 500 MHz. Since the frequency is 500 times larger than in Example 5.1-2, the induced voltage will be 500 times higher as 435 V; the smaller dimension of the loop is 0.083λ. What a difference the frequency has made!

Example 5.1-4. A uniform magnetic flux density of 1 wb/m^2 operates at a frequency of 60 Hz. Calculate the electric field intensity. For convenience, assume that the magnetic flux is in the $\mathbf{a}_Z$ direction. We know that a voltage will be induced in a wire loop located in the $z = 0$ plane, so there must be an E_ϕ. From the differential form of Faraday's law, $\nabla \times \mathbf{E} = \mathbf{a}_Z \frac{\partial(\rho E_\phi)}{\rho\partial\rho} = -\frac{\partial \mathbf{B}}{\partial t} = -\mathbf{a}_Z 377 \cos 377t$. Equating the second and fourth terms and integrating, we find $\mathbf{E} = -\mathbf{a}_\phi 188.5\rho \cos 377t$ V/m. Since $\mathbf{E}$ is uniform, the voltage induced in a circular loop of radius ρ is readily calculated as $V_{\text{LOOP}} = E_\phi 2\pi\rho = -377\pi\rho^2 \cos 377t$ V. An alternate calculation uses the integral form of Faraday's law $V_{\text{LOOP}} = -\pi\rho^2 d(\sin 377t)/dt = -377\pi\rho^2 \cos 377t$ V as before.

5.2 AMPERE'S LAW REVISITED

Since you now know that in Chapters 1 and 2 you were led to falsely believe that all electric fields are conservative, it's time that you know that there is more to Ampere's law as well. The situation is best viewed in Fig. 5.2 where the work integral of the magnetic field intensity is evaluated on the path L.

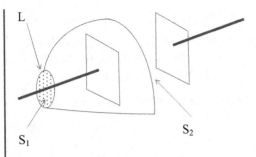

S_1

FIGURE 5.2: Ampere's law.

According to the earlier work in Chapter 3, this integral is equal to the current enclosed as

$$\oint_L \mathbf{H} \cdot \mathbf{dl} = I_{\mathrm{ENC}}. \qquad (5.8)$$

The enclosed current can be evaluated by calculating the current crossing any surface the edges of which are the path L. When the surface S_1 is chosen, the current which flows in the PEC wire lead of the capacitor is the capacitor current I_C. A more accurate description includes a current density, $\mathbf{J}$, distributed throughout the wire. With the path L snugly fitted around the outside of the wire, Eq. (5.8) becomes

$$\oint_L \mathbf{H} \cdot \mathbf{dl} = I_{\mathrm{ENC}} = \iint_{S_1} \mathbf{J} \cdot \mathbf{ds} \qquad (5.9)$$

as we learned in Chapter 3. Now consider the current through surface S_2 that encloses just one of the capacitor electrodes. The LHS of Eq. (5.8) still has the same value as before since the path L has not changed. But, there is no electric current flowing through S_2 and the RHS is apparently zero. How can this contradiction be resolved? Fortunately for us, James Clerk Maxwell proposed a solution to this problem nearly 150 years ago. His reasoning was based on a mechanical model with rotating gears in contrast to the model of fields in space. His reasoning in the electromagnetic domain went somewhat in the following manner.

There must be a more general description of current that penetrates surface S_2. Recall that in a parallel-electrode capacitor, current flows into the capacitor and deposits a uniform surface charge density on the electrodes given by

$$\rho_S(t) = \frac{Q(t)}{A} = \frac{\displaystyle\int_{t'=t_o}^{t} I_C(t')dt'}{A}. \qquad (5.10)$$

The electric flux density between the electrodes is expressed by

$$D_n(t) = \rho_S(t) = \frac{\displaystyle\int_{t'=t_o}^{t} I_C(t')dt'}{A}. \qquad (5.11)$$

A corresponding electric flux crosses the surface S_2

$$\Psi_{e\,S_2} = \iint_{S_2} \mathbf{D} \cdot \mathbf{ds} = \frac{\int_{t'=t_o}^{t} I_C(t')\,dt'}{A} A = \int_{t'=t_o}^{t} I_C(t')\,dt' \tag{5.12}$$

where the integration is simplified to a multiplication due to the uniform charge density. The RHS of Eq. (5.11) would give the same result for nonuniform charge density; the calculation would be more involved. The time derivative of the electric flux becomes

$$\frac{\partial \Psi_{e\,S_2}}{\partial t} = \iint_{S_2} \frac{\partial \mathbf{D}}{\partial t} \cdot \mathbf{ds} = \frac{\partial}{\partial t} \int_{t'=t_o}^{t} I_C(t')\,dt'$$

$$= I_C(t) \tag{5.13}$$

where the surface S_2 is unchanging with time. This is exactly the value obtained by integration over surface S_1 where there was only the electric current density. On surface S_2 where there is no electric current present the same result is obtained by using this new term $\partial \Psi_e / \partial t$. Recall that electric flux density has units of Coulombs/meter^2 so that the expression is dimensionally correct, Coulombs/second/meter^2. With this information, Maxwell proposed that Ampere's law could be made correct for all applications if it were modified to include this new term. Specifically, he defined $\partial \mathbf{D} / \partial t$ as *displacement current* in contrast to $\mathbf{J}$ known as *conduction current*. The electric charges on one electrode where the electric flux lines begin displace charges on another surface where the electric flux lines ended. He stated that Ampere's law is correct when the sum of both conduction and displacement current is included as

$$\oint_L \mathbf{H} \cdot \mathbf{dl} = I_{ENC} = \iint_S \left(\mathbf{J} + \frac{\partial \mathbf{D}}{\partial t} \right) \cdot \mathbf{ds}. \tag{5.14}$$

For the capacitor, surface S_1 contains only conduction current while surface S_2 contains only displacement current. For DC and low frequencies where the displacement current outside the capacitor electrodes is negligible, the results are identical to our earlier form as they must be.

Using similar mathematical steps with Ampere's law as we did with Faraday's law in Eqs. (5.4)–(5.7), we obtain the differential form of Ampere's law as

$$\nabla \times \mathbf{H} = \mathbf{J} + \frac{\partial \mathbf{D}}{\partial t}. \tag{5.15}$$

The magnetic field at a point is proportional to the sum of the conduction and displacement currents. Moreover, it is perpendicular to their sum.

A final aspect of the differential form of Ampere's law comes from vector identities that show that the divergence of the curl of a vector is identically zero. Applying this to Eq. (5.15), we have

$$\nabla \cdot \nabla \times \mathbf{H} = \nabla \cdot \left(\mathbf{J} + \frac{\partial \mathbf{D}}{\partial t} \right) = \nabla \cdot \mathbf{J} + \frac{\partial \nabla \cdot \mathbf{D}}{\partial t}$$

$$= \nabla \cdot \mathbf{J} + \frac{\partial \rho_V}{\partial t} \equiv 0 \tag{5.16}$$

where the point form of Gauss' law, $\nabla \cdot \mathbf{D} = \rho_V$, has been used. Since the last equality is just the point form of the conservation of charge, this form of Ampere's law is consistent with the physical reality of charge conservation.

Example 5.2-1. A current of 100 mA flows in a copper wire of 1 mm radius. Calculate the conduction and displacement current densities within the wire for an operating frequency of 60 Hz. Assuming a uniform current density within the wire, $|\mathbf{J}| = 0.1/\pi(0.001)^2 = 31.8$ kA/m^2. Within the copper wire, $|\mathbf{E}| = 31.8 \times 10^3/5.8 \times 10^7 = 0.549$ mV/m; $|\mathbf{D}| = \varepsilon_o(0.549 \times 10^{-3}) = 4.9 \times 10^{15}$ C/m^2. To calculate the displacement current magnitude within the wire, we multiply by ω. For 60 Hz, this gives $|\partial \mathbf{D}/\partial t| = 1.83$ pA/m$^2 \ll 31.8$ kA/m^2. We certainly can neglect displacement current in copper wires for low frequency operation.

5.3   MAXWELL'S EQUATIONS

The set of equations that govern the behavior of time-varying electromagnetic fields have been collectively named *Maxwell's equations*. They are named in his honor primarily because the generalizations that he made to Ampere's law made the set of equations consistent and valid for all circumstances. Table 5.1 contains the time domain form of Maxwell's equations in both integral and differential forms. In addition to the presence of time derivatives, the time domain form of the fields is emphasized by the script form just as we do in circuits. In addition, the field quantities have a spatial dependence, i.e., $\mathcal{E} = \mathcal{E}(\mathbf{r}, t)$. For convenience, we write the field quantities without the functional dependence as $\mathcal{E}$ except in cases where it is important to emphasize it.

As in circuits, a great deal of information can be obtained from studying the solution to these equations when the sources are of sinusoidal or *time-harmonic* form. Many engineering applications involve a single frequency or a narrow band of frequencies that can be treated as a single frequency. As in circuits, the phasor solutions can be combined with Fourier techniques to calculate the response to sources with arbitrary waveshapes. As you recall, phasor domain

TABLE 5.1: The Maxwell's Time-domain Equations.

INTEGRAL FORM	DIFFERENTIAL FORM
$\oint_L \boldsymbol{\mathcal{E}} \cdot \mathbf{dl} = -\iint_S \frac{\partial \boldsymbol{\mathcal{B}}}{\partial t} \cdot \mathbf{ds}$	$\nabla \times \boldsymbol{\mathcal{E}} = -\frac{\partial \boldsymbol{\mathcal{B}}}{\partial t}$
$\oint_L \boldsymbol{\mathcal{H}} \cdot \mathbf{dl} = \iint_S \left(\boldsymbol{\mathcal{J}} + \frac{\partial \boldsymbol{\mathcal{D}}}{\partial t} \right) \cdot \mathbf{ds}$	$\nabla \times \boldsymbol{\mathcal{H}} = \boldsymbol{\mathcal{J}} + \frac{\partial \boldsymbol{\mathcal{D}}}{\partial t}$
$\oiint_S \boldsymbol{\mathcal{D}} \cdot \mathbf{ds} = Q_{\text{ENC}}$	$\nabla \cdot \boldsymbol{\mathcal{D}} = \rho_V$
$\oiint_S \boldsymbol{\mathcal{B}} \cdot \mathbf{ds} = 0$	$\nabla \cdot \boldsymbol{\mathcal{B}} = 0$
$\oiint_S \boldsymbol{\mathcal{J}} \cdot \mathbf{ds} = -\frac{dQ}{dt}$	$\nabla \cdot \boldsymbol{\mathcal{J}} = -\frac{\partial \rho_V}{\partial t}$

forms of voltage and current are obtained as the real part of a rotating phasor $v(t) = Re\{V_P e^{j\omega t}\}$ where V_P is a time-independent, complex phasor. The same procedure is applied to each of the vector components of an electromagnetic field where $\boldsymbol{\mathcal{E}}_i(\mathbf{r}, t) = Re\{E_i(\mathbf{r})e^{j\omega t}\}$. $E_i(\mathbf{r})$ represents ith component of a complex phasor in one of our three favorite coordinate systems, (x, y, z), (ρ, ϕ, z), or (r, θ, ϕ). Therefore, the phasor form of the electric field vector in Cartesian coordinates is

$$\begin{aligned} \mathbf{E}(\mathbf{r}) &= \mathbf{a}_X E_X e^{j\phi_X} + \mathbf{a}_Y E_Y e^{j\phi_Y} + \mathbf{a}_Z E_Z e^{j\phi_Z} \\ &= \mathbf{a}_X (E_{Xr} + j E_{Xi}) + \mathbf{a}_Y (E_{Yr} + j E_{Yi}) \\ &\quad + \mathbf{a}_Z (E_{Zr} + j E_{Zi}). \end{aligned} \tag{5.17}$$

The phasor domain form of a vector field is represented by six real values. In addition, each of these values can have a spatial dependence as well, e.g., $E_{Xr} = E_{Xr}(x, y, z)$. Maxwell's equations in phasor form are similar to the time domain form, but due to the sinusoidal time behavior, all time derivatives are replaced by $j\omega$. The phasor domain form of Maxwell's equations is shown in Table 5.2. In the time domain, each field symbol represents three real vector components with spatial and time dependence. In the phasor domain, they represent three complex vector components with an implied time dependence of $e^{j\omega t}$.

With the addition of time dependence, the electric and magnetic fields are inseparable. They occur together and are connected through the first two of Maxwell's equations. In both equations, the work integral of one field intensity is due to the time variation of the flux of the other field enclosed by the path of the work integral. The fields can be viewed as interlinked

TABLE 5.2: Maxwell's Phasor Domain Equations.

INTEGRAL FORM	DIFFERENTIAL FORM
$\oint_L \mathbf{E} \cdot \mathbf{dl} = -j\omega \iint_S \mathbf{B} \cdot \mathbf{ds}$	$\nabla \times \mathbf{E} = -j\omega\mathbf{B}$
$\oint_L \mathbf{H} \cdot \mathbf{dl} = \iint_S (\mathbf{J} + j\omega\mathbf{D}) \cdot \mathbf{ds}$	$\nabla \times \mathbf{H} = \mathbf{J} + j\omega\mathbf{D}$
$\oiint_S \mathbf{D} \cdot \mathbf{ds} = Q_{\text{ENC}}$	$\nabla \cdot \mathbf{D} = \rho_V$
$\oiint_S \mathbf{B} \cdot \mathbf{ds} = 0$	$\nabla \cdot \mathbf{B} = 0$
$\oiint_S \mathbf{J} \cdot \mathbf{ds} = -j\omega Q$	$\nabla \cdot \mathbf{J} = -j\omega\rho_V$

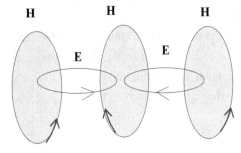

FIGURE 5.3: Linking of time-varying electric and magnetic fields.

one with another as shown in Fig. 5.3. Time variations in the electric flux density produce a magnetic field intensity; time variations in the magnetic flux density produce an electric field intensity. One time-varying field cannot exist without the other. This inextricable linking of the electric and magnetic fields has caused some to refer to them as a single *electromagnetic field*.

The interlinked electric and magnetic fields will lead us to quite different behavior than we observed in lumped elements. The next sections will reveal some exciting new properties of lumped elements.

5.4   QUASI-STATIC ANALYSIS OF LUMPED ELEMENTS

What is the evidence that the elements defined for electro- and magnetostatics are valid for time-varying signals? Most often the DC parameters are used for AC calculations, but without any investigation into the range of their validity. In fact, the following work will show that modifications of the DC models are vital to accurate models. But, never fear, these modifications are readily obtained from basic electromagnetic principles. The following material will use these principles to develop the AC models.

This development will use a relatively simple geometry and will impose a number of assumptions that will make the mathematics simpler, but will not lose the essential features of

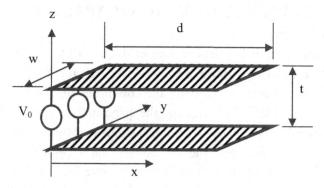

FIGURE 5.4: Lumped element capacitor.

high frequency, lumped element modifications. Several assumptions will hold for the lumped element development and discussion.

The lumped elements are comprised of planar electrodes arranged in a parallel configuration and composed of PEC material. The electrodes are of length d, width w, and separation t with the constraints of $d \geq w \gg t$. Consequently, there are no fringing fields. The inter-electrode region of the lumped element is filled with homogeneous material.

Time-harmonic AC excitation of the elements is applied by uniformly distributed sources along the left edge of the element with uniform voltage, current, charge, and fields in the y-direction, i.e., $\partial/\partial y = 0$. These assumptions are shown in Fig. 5.4 for a capacitor; the inductor and resistor are similar.

With this set of assumptions in place, let's outline the process use to develop the AC model for circuit elements. The approach will be to expand the voltages in a power series with respect to frequency of the form

$$V(\omega) = V_0 + \omega V_1 + \omega^2 V_2 + \cdots \tag{5.18}$$

Moreover, currents, charges, and fields can be represented in similar power series in ω. In this approach, a voltage/current/charge/field quantity for the nth order term becomes the source for the next higher order, $(n + 1)$st term through the connections provided by time derivatives, i.e., $j\omega$ in the phasor domain. For example, the zero order voltage, V_0, and the zero order capacitance, C_0, are calculated from static principles. Using V_0 as the source along with C_0, the first order current I_1 can be calculated. This process is valid for calculating the associated fields within the capacitor as well. For example, since V_0 is used to calculate E_0 it has a finite value. There is no I_0 so there is no H_0. But, there is a finite I_1 so there is a finite H_1. Moreover, the process can be extended to higher order terms. The quasi-static models for three circuit elements—capacitors, inductors, and resistors—will be calculated. This calculation process involves a mixture of circuit and field concepts.

5.5   OPEN-CIRCUIT TERMINATION: CAPACITORS

5.5.1   Zero Order Term

A uniform DC voltage is applied at the left edge of the two PEC electrodes of the capacitor shown at right. A dielectric with permittivity of ε fills the region between the electrodes. With the configuration of the electrodes satisfying the conditions of $d \geq w \gg t$, the zero order charge distribution on the PEC electrodes and the associated zero order electric field within the electrode are essentially uniform. Furthermore, assume that there is no fringing of the field at the edges of the capacitor. The zero order DC voltage, V_0, is applied to the electrodes via the sources along the left edge of the electrodes. There is a surface charge density, ρ_{S0}, distributed uniformly on the inner, facing surfaces of the PEC electrodes. The boundary condition at the PEC electrode requires that the normal component of the zero order electric flux density is equal to the zero order surface charge density, i.e.,

$$D_{n0} = \rho_{S0}, \tag{5.19}$$

and is directed downward in the negative z-direction from positive charge on the top electrode to the corresponding negative charge on the bottom electrode.

The relationship of $\mathbf{D} = \varepsilon\mathbf{E}$ and the well-known relationship between the electrode voltage and the uniform electric field of the parallel electrode capacitor gives

$$\rho_{S0} = \varepsilon E_{n0} = \varepsilon\frac{V_0}{t}. \tag{5.20}$$

The total charge on the upper electrode is expressed as

$$Q_0 = \iint\limits_{\text{Electrode}} \rho_{S0}\, da = \int_{y=0}^{w}\int_{x=0}^{d} \rho_{S0}\, dx dy$$

$$= \int_{y=0}^{w}\int_{x=0}^{d} \varepsilon\frac{V_0}{t} dx dy = \varepsilon\frac{V_0}{t}wd. \tag{5.21}$$

The definition of capacitance $C = Q/V$ leads to the well-known DC, or zero order, capacitance of

$$C_{\text{DC}} = C_0 = \frac{Q_0}{V_0} = \frac{\varepsilon\dfrac{V_0}{t}wd}{V_0} = \frac{\varepsilon wd}{t}. \tag{5.22}$$

Since there is no zero order current flow, i.e., $I_0 = 0$, the zero order magnetic field is zero, $H_0 = 0$.

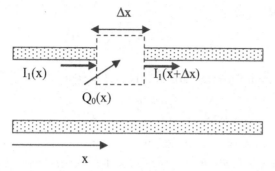

FIGURE 5.5: Conservation of charge on an electrode.

5.5.2   First Order Term

So what happens when the applied voltage is time varying? Intuitively, we know that time-varying electric fields within the capacitor will produce a magnetic field. Alternatively, from a circuit point of view, a time-varying voltage impressed on a capacitor will produce a time-varying conduction current entering the terminals of the capacitor and flowing on the surface of the electrodes. But what are the features of this current? A fundamental description of the current that flows on the PEC electrodes is pictured in Fig. 5.5.

Charge conservation requires that the net current out of any region is equal to the negative time rate of change of charge within the region. The current flowing into the dashed region at position x on the upper PEC from the left is $I_1(x)$; the current leaving the region on the right is $I_1(x + \Delta x)$; the net current out of the region is $\Delta I_1(x) = I_1(x + \Delta x) - I_1(x)$. The charge contained within the incremental region Δx is $\Delta Q_0(x) = \rho_{s0}w\Delta x$; the rate of change of charge within the region is $j\omega Q_0(x) = j\omega\rho_{s0}w\Delta x$. Combining these two terms leads to

$$\underbrace{\Delta I_1(x) = I_1(x + \Delta x) - I_1(x) = -j\omega\rho_{s0}w\Delta x}$$
$$\Downarrow$$
$$\frac{I_1(x + \Delta x) - I_1(x)}{\Delta x} \doteq \frac{dI_1(x)}{dx} = -j\omega\rho_{s0}w, \qquad (5.23)$$

a differential equation that describes the first order current into the capacitor terminals in terms of the zero order charge calculated previously. This equation can be integrated and evaluated by noting that there is no current flow at the right edge of the PEC so that $I(d) = 0$. The equation for the first order current becomes

$$I_1(x) = j\omega\rho_{s0}w\,(d - x). \qquad (5.24)$$

This can be expressed in terms of the zero order electric field as

$$I_1(x) = j\omega\varepsilon\,E_0 w\,(d - x) \qquad (5.25)$$

or in terms of the zero order voltage

$$I_1(x) = \frac{j\omega V_0 w\varepsilon(d-x)}{t}.$$ (5.26)

The input first order current into the capacitor is at $x = 0$ or

$$I_{1IN}(0) = \frac{j\omega V_0 w\varepsilon d}{t} = j\omega V_0 C_0.$$ (5.27)

This first order input current is of the form used in circuits to express the voltage–current ratio for capacitors in phasor form. From this form, the first order circuit model is calculated as the ratio of the sum of preceding of orders of voltage and current and is expressed as

$$Z_1 = \frac{V_0}{I_{1IN}(0)} = \frac{V_0}{\frac{j\omega V_0 w\varepsilon d}{t}} = \frac{1}{j\omega C_0} = \frac{1}{j\omega C_{DC}}.$$ (5.28)

The first order circuit model is merely the C_{DC} for time-varying signals. So we can use the static capacitance calculations for time-varying signals; this is called the quasi-static model. But, what is the upper limit of frequency at which we can use this model? That is explained in the next section.

5.5.3   Second Order Term

The current $I_1(x)$ which flows from left to right on the upper electrode (and oppositely on the lower electrode) is uniform from front to back on the electrode; since this current flows only on the inner surfaces of the electrode, the surface current density is expressed as

$$K_1(x) = \frac{I_1(x)}{w}.$$ (5.29)

But the current flow is not uniform along the electrode in the x-direction. This can be explained and visualized with the help of Fig. 5.6. All of the uniformly distributed, zero order charge flows from the upper to the lower electrode twice every cycle, alternatively positive, then negative on the upper electrode. The surface current density at the left edge of the electrodes is supplying all

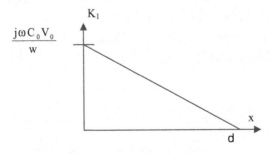

FIGURE 5.6: Conservation of charge and current on electrodes.

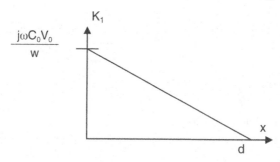

FIGURE 5.7: Linear variation of a surface current, K_1.

of the charge on the electrode, but at locations to the right, there is less charge that the current must supply. As the observation point is located farther to the right, less charge is present on the electrode. Each arrow represents a unit of current moving to or from the left edge of the electrode. The number of arrows at a location represents the strength of the current at that point. At the right edge, the current is zero; at the left edge, the current has a maximum. The variation from one edge to the other is a linear decrease from the left to the right as shown by the linear dependence of $I_1(x)$ in Fig. 5.7. The surface current density is expressed as

$$K_1(x) = \frac{j\omega C_0 V_0}{w}\frac{(d-x)}{d}. \tag{5.30}$$

From the first order surface current density, the magnetic field intensity using the well-known boundary condition for tangential components of the magnetic field is expressed as

$$\mathbf{a}_{nA} \times (\mathbf{H}_A - \mathbf{H}_B) = \mathbf{K}. \tag{5.31}$$

In the vicinity of the upper electrode, this sets up an H-field both above and below the surface that has a magnitude of $K/2$; below the upper electrodes, the magnetic field is directed into the plane of the paper, while above the upper electrode, it is directed out of the plane of the paper. In a similar fashion, the oppositely directed surface current on the lower electrode sets up a magnetic field into the plane of the paper above the lower electrode and out of the plane of the paper below the lower electrode. Since it is a linear system, superposition applies and we can add the two fields to get the total field. The two contributions add in the inter-electrode region; they cancel in the region outside the electrodes as shown in Fig. 5.8. Consequently, the magnetic field points into the plane of the paper and is calculated as

$$H_1(x) = K_1(x) = \frac{I_1(x)}{w}. \tag{5.32}$$

Alternatively, the magnetic field intensity can be found by the application of Ampere's law on path L_A that encircles the upper electrode and is confined to a path in a constant x-plane; see Fig. 5.9. The only current enclosed by this path is the surface current density on

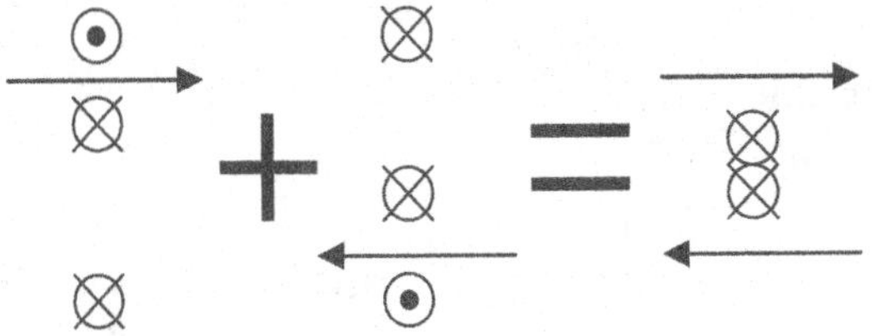

FIGURE 5.8: Magnetic field due to a surface current.

the electrode. This leads to

$$\oint_{L_A} \mathbf{H}_1(x) \cdot \mathbf{dl} = H_1(x)w = I_1(x) \tag{5.33}$$

so that

$$H_1(x) = \frac{I_1(x)}{w} = K_1(x) = \frac{j\omega C_0 V_0}{w}\frac{(d-x)}{d}$$
$$= \frac{j\omega V_0 \varepsilon (d-x)}{t} \tag{5.36}$$

as before.

A second alternative is to use Ampere's law in the inter-electrode region

$$\oint_{L_B} \mathbf{H}_1 \cdot \mathbf{dl} = j\omega\Psi_{e0} = j\omega \iint_{S_B} \mathbf{D}_0 \cdot \mathbf{ds} \tag{5.37}$$

where there is no conduction current, only displacement current. The path L_B (see Fig. 5.9) is confined to a horizontal planar S_B (located, for example, midway between the two electrodes at $z = t/2$) through which the zero order displacement current passes. The front, right, and back edges of the surface S_B are aligned with the corresponding edges of the PEC electrodes. The left edge of the planar surface S_B is located at position x. These edges define the path of

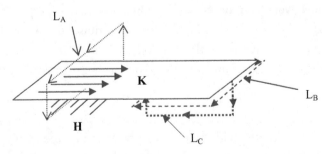

FIGURE 5.9: Integration paths for Eqs. (1.34) and (1.35).

L_B. Since the zero order electric flux density is uniform, the total displacement current can be expressed as

$$j\omega\Psi_{e0} = j\omega \int_{x'=x}^{d} \int_{y'=0}^{w} D_0 \, dy dx' = j\omega\rho_{s0}w \, (d - x). \tag{5.38}$$

The first order magnetic field is integrated along the path L_B that encloses the surface S_B with the path clockwise when viewed in the negative z-direction from above (due to the right-hand relationship between the positive direction chosen for the downward electric flux and the enclosing path). Since the zero order magnetic field intensity is perpendicular to the path L_B on the front and back segments of the path, the integral vanishes along these two segments. Since the zero order current density (and the corresponding magnetic field intensity) is zero at the right end of the electrodes, the segment of the integral at this position vanishes as well. Finally, the first order magnetic field along the remaining segment of the line integral is uniform and in the direction of integration the integral along this segment is just the magnetic field multiplied by the width. This is expressed mathematically as

$$\oint_{L_B} \mathbf{H}_1 \cdot \mathbf{dl} = H_1(x)w. \tag{5.39}$$

Equating the two previous equations, we can solve for the zero order magnetic field intensity as

$$H_1(x) = j\omega\rho_{s0} \, (d - x) = \frac{j\omega V_0 \varepsilon \, (d - x)}{t}$$
$$= j\omega\varepsilon \, V_0 C_0 \frac{(d - x)}{wd} \tag{5.40}$$

as calculated earlier. The magnetic flux density is calculated as

$$B_1(x) = \mu H_1(x) = \frac{\mu I_1(x)}{w} = \frac{j\omega\mu C_0 V_0 \, (d - x)}{w \quad d}. \tag{5.41}$$

This field is directed in the y-direction, into the plane of the paper.

Of course, this first order, time-varying magnetic field will induce a second order electric field. This electric field can be calculated according to Faraday's law as

$$\oint_{L_C} \mathbf{E}_2 \cdot \mathbf{dl} = -j\omega \iint_{S_C} \mathbf{B}_1 \cdot \mathbf{ds} \tag{5.42}$$

The surface S_C is a vertical surface (located for convenience along the mid-plane between the two electrodes, $y = w/2$) extending axially from the right end of the electrodes, $x = d$, to the arbitrary position x. The path L_C encloses the surface area S_C according to the right-hand rule based upon the first order magnetic field directed into the plane of the paper; the path L_C is directed CW when looking in the y-direction.

In considering the left-hand side of the previous equation, note that the upper and lower horizontal segments of the path are evaluated along the PEC electrodes where electric fields are zero so there is no contribution to the integral by these segments. The right-hand vertical path segment is located at the end of the electrodes where the current and H-field are zero. It is plausible that the induced voltage will be zero at this location and, hence, so will the electric field. Consequently, this segment will contribute zero to the integral. The only segment of the path that contributes to the integral is along the left hand, vertical segment of the path so that the integral is evaluated as

$$\oint_{L_C} \mathbf{E}_2 \cdot \mathbf{dl} = E_2(x)t. \tag{5.43}$$

Note that this segment is directed upward and the integral will be positive for E_2 directed upward in the z-direction as well.

Evaluation of the integral on the right-hand side of the equation can be accomplished readily using the first order magnetic field as

$$-j\omega \iint_{S_C} \mathbf{B}_1 \cdot \mathbf{ds} = -j\omega \int_{x'=x}^{d} \int_{z=0}^{t} \mu H_1(x') \, dy dx'$$

$$= \frac{-j\omega t\mu}{w} \int_{x'=x}^{d} I_1(x') \, dx'$$

$$= -\frac{(j\omega)j\omega\mu t C_0 V_0}{w} \int_{x'=x}^{d} \frac{(d-x')}{d} \, dx'$$

$$= \frac{\omega^2 \mu t C_0 V_0}{2wd} (d-x)^2 . \tag{5.44}$$

Combining the results of the previous two equations leads to

$$E_2(x) = \frac{\omega^2 \mu C_0 V_0}{2wd} (d-x)^2 . \tag{5.45}$$

The positive sign indicates that the second order field is directed in the positive z-direction and opposes the zero order field. The corresponding second order voltage at the left end of the electrodes is

$$V_2(x) = -\int_{z=0}^{t} \frac{\omega^2 \mu C_0 V_0}{2wd}(d-x)^2\, dz$$

$$= -\frac{\omega^2 \mu t C_0 V_0}{2wd}(d-x)^2 = -\frac{\omega^2 \mu \varepsilon\,(d-x)^2}{2}V_0. \qquad (5.46)$$

Since the first order electric field is zero, the second order current and magnetic fields (which depend upon the first order electric field) are zero or $I_2 = H_2 = 0$.

The second order impedance can be expressed as the quotient of the total voltage and the total current (both correct to the second order). The input voltage (at $x = 0$) correct to the second order is expressed as

$$V_{0,2}(0) = V_0 + \omega^2 V_2 = V_0 - \frac{\omega^2 \varepsilon \mu d^2}{2}V_0$$

$$= V_0\left(1 - \frac{\omega^2 \varepsilon \mu d^2}{2}\right); \qquad (5.47)$$

the input current correct to the second order is expressed as

$$I_1(0) = \frac{j\omega V_0 w \varepsilon d}{t}. \qquad (5.48)$$

This leads to the second order input impedance as

$$Z_2 = \frac{V_{0,2}(0)}{I_1(0)} = \frac{V_0\left(1 - \dfrac{\omega^2 \varepsilon \mu d^2}{2}\right)}{\dfrac{j\omega V_0 w \varepsilon d}{t}}$$

$$= \frac{1}{j\omega\dfrac{\varepsilon w d}{t}} + \frac{-\dfrac{\omega^2 \varepsilon \mu d^2}{2}}{j\omega\dfrac{\varepsilon w d}{t}} = \frac{1}{j\omega C_0} + j\omega\frac{\mu d t}{2w}$$

$$= \frac{1}{j\omega C_0} + j\omega\frac{L_0}{2}, \qquad (5.49)$$

the series connection of the quasi-static capacitance C_0 in series with half the DC inductance of this structure, $L_0/2$, as represented by the circuit shown in Fig. 5.10. The "half" inductance is due to the linear variation of the magnetic field from zero to the peak value along the length. This gives a total magnetic flux enclosed that is half the magnetic flux for a uniform magnetic field, i.e., the field of zero order, DC magnetic fields.

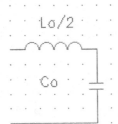

FIGURE 5.10: Second order model for a capacitor.

Now we can get an idea of the upper frequency limit at which the DC capacitance, $C_{DC} = C_0$, accurately represents the behavior of a capacitor. For frequencies where the second term of the voltage expression is negligible, i.e.,

$$\frac{\omega^2 \varepsilon \mu d^2}{2} = C_0 L_1 = \frac{C_0 L_0}{2} \ll 1, \qquad (5.50)$$

the DC capacitance can be used. The absolute upper frequency limit of operation of the element as a capacitor occurs when the input impedance shows a series resonance with an input impedance of a short circuit, i.e., the inductive reactance is equal to the capacitive reactance, or

$$\underbrace{\omega L_1 = \frac{1}{\omega C_0}}$$

$$\Downarrow$$

$$\omega_{Res} = \frac{1}{\sqrt{L_1 C_0}} = \frac{\sqrt{2}}{\sqrt{\mu \varepsilon} d}$$

$$= \frac{v_p \sqrt{2}}{d}. \qquad (5.51)$$

At DC, the input impedance is an open circuit; at this resonant frequency, the input impedance is a short circuit!! The length in the x-direction at which this resonance occurs can be found by solving for d as

$$d_{reseonance} = \frac{\sqrt{2}}{2\pi} \lambda \approx 0.225\lambda. \qquad (5.52)$$

This is the length of the capacitor in which the quasi-static model predicts that there is as much magnetic energy stored in the inductor as the electric energy stored in the capacitor. It is manifest as a series resonant circuit with a short-circuit input impedance. The element has ceased to be a capacitor, and at still higher frequencies it appears as an inductor! The highest frequency of operation as a capacitor must be significantly below this limit.

5.5.4 Higher Order Terms

This process can be continued to higher order terms. The second order voltage has a surface charge on the electrodes associated with it and in turn an associated second order electric flux density. The time-varying second order electric flux induces a third order magnetic field intensity and a third order current on the electrodes. And so on ad infinitum. The form of the higher order terms follows a definite pattern.

The third order input current is given by

$$I_3(0) = j\omega C_2 V_2(0) = j\omega \frac{\varepsilon wd}{3t} V_2$$

$$= -j\omega \frac{\varepsilon wd}{3t} \frac{\omega^2 \varepsilon \mu d^2}{2} V_0$$

$$= -j\omega^3 \frac{\mu \varepsilon^2 wd^3}{6t} V_0. \qquad (5.53)$$

This leads to the third order impedance is then expressed as

$$Z_3 = \frac{V_0 + \omega^2 V_2}{I_1 + \omega^2 I_3}$$

$$= \frac{\left(1 - \dfrac{\omega^2 \varepsilon \mu d^2}{2}\right) V_0}{j\omega \dfrac{\varepsilon wd}{t} \left(1 - \omega^2 \dfrac{\mu \varepsilon d^2}{6}\right) V_0}$$

$$= \frac{\left(1 - \dfrac{\omega^2 \varepsilon \mu d^2}{2}\right)}{j\omega C_0 \left(1 - \omega^2 \dfrac{\mu \varepsilon d^2}{6}\right)}. \qquad (5.54)$$

The expression for the impedance as shown in Z_3 is unique. However, there are many forms of realization. For example, the Foster type I method of synthesis leads to a manipulation of the Z_3 equation to the form

$$Z_3 = \frac{\left(1 - \dfrac{\omega^2 \varepsilon \mu d^2}{2}\right)}{j\omega C_0 \left(1 - \omega^2 \dfrac{\mu \varepsilon d^2}{6}\right)}$$

$$= \frac{1}{j\omega C_0} + \frac{j\omega \left(\dfrac{L_0}{3}\right) \dfrac{1}{j\omega \left(\dfrac{C_0}{2}\right)}}{j\omega \left(\dfrac{L_0}{3}\right) + \dfrac{1}{j\omega \left(\dfrac{C_0}{2}\right)}}, \qquad (5.55)$$

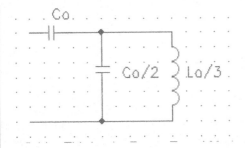

FIGURE 5.11: Third order Foster type I model for a lumped capacitor.

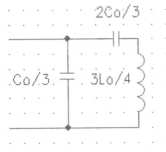

FIGURE 5.12: Third order Cauer type II model for a lumped capacitor.

representing a capacitor of C_0 in series with the parallel combination of $L_0/3$ and $C_0/2$; see Fig. 5.11.

Alternatively, the Cauer type II method leads to the realization of a shunt connection of a capacitor of $C_0/3$ in parallel with the series connection of an inductor of $3L_0/4$ and a capacitor of $2C_0/3$, see Fig. 5.12.

The implementation of the first two orders of impedance leads to a unique configuration, but for all higher orders the implementation of the impedances is not unique.

The terms in the numerator of the impedance expression (that represents the input voltage on the line) obey a pattern. With the inclusion of all of the higher order terms, the voltage can be put in a closed form as

$$V_{\text{Terminals}}$$
$$= V_0 \left(1 - \frac{\omega^2 \mu \varepsilon d^2}{2!} + \frac{\left(\omega\sqrt{\mu\varepsilon}d\right)^4}{4!} - \frac{\left(\omega\sqrt{\mu\varepsilon}d\right)^6}{6!} - \cdots \right)$$
$$= V_0 \cos\left(\omega\sqrt{\mu\varepsilon}d\right) = V_0 \cos(kd) \tag{5.56}$$

where k will be defined and explained later as $k = \omega\sqrt{\mu\varepsilon} = 2\pi/\lambda$. As we shall see later, this is the closed form expression for the voltage on an open-circuited transmission line, a case where $d \gg w$. This closed form expression is the exact wave solution, well beyond the quasi-static frequency range for a capacitor. An examination of the closed form, exact representation reveals that the input voltage is zero, i.e., $\cos(kd) = 0$, when the length $d = \lambda/4$ which compares very well with the two-element, second order impedance approximation that predicts a zero value for a value of $d = 0.225\lambda$ as predicted by Eq. (5.57). This should give some confidence in the accuracy of the quasi-static solutions.

In a similar manner, the terms in the denominator of the impedance expression (that represents the input current on the line) lead to a closed form as

$$I_{\text{Terminals}} = I_1 \left(\omega\sqrt{\mu\varepsilon}d - \frac{\left(\omega\sqrt{\mu\varepsilon}d\right)^3}{3!} + \frac{\left(\omega\sqrt{\mu\varepsilon}d\right)^5}{5!} - \cdots \right)$$
$$= I_1 \sin\left(\omega\sqrt{\mu\varepsilon}d\right) = I_1 \sin(kd). \tag{5.58}$$

This means that the input impedance of open-circuited electrodes is expressed as

$$Z_{\text{INOC}} = \frac{V_{\text{Terminals}}}{I_{\text{Terminals}}} = \frac{V_0}{I_1} \cot(kd) = \frac{V_0}{I_1} \cot(\omega\sqrt{\mu\varepsilon}\,d). \tag{5.59}$$

This is quite a change from the rather simple, DC, and low-frequency response of a lumped element capacitor. Such a dramatic behavior may not be limited to capacitors; let's look at the behavior of lumped inductors at higher frequencies.

5.6 SHORT-CIRCUIT TERMINATION: INDUCTORS

5.6.1 Zero Order Term

A uniform current is applied to the left edge of the two PEC electrodes of the inductor shown in Fig. 5.13. Magnetic material with permeability of μ fills the region between the electrodes. With this configuration of the electrodes satisfying the conditions of $d \geq w \gg t$, the surface current distribution and the magnetic field within the electrodes are essentially uniform in the transverse direction, i.e., the y direction. Furthermore, assume that there is no fringing of the field at the front and back edges of the inductor. For these conditions, consider that a DC current, I_o, is applied to the electrodes via equal current sources along the left edge of the electrodes.

There is a surface current density, K_0, distributed uniformly on the inner facing surfaces of the PEC electrodes. As calculated earlier, the zero order magnetic field within the inductor is directed into the plane of the paper and is equal to the zero order surface current density as

$$H_0 = K_0 = \frac{I_0}{w} \tag{5.60}$$

and the zero order magnetic flux density is expressed as

$$B_0 = \mu H_0 = \frac{\mu I_0}{w}. \tag{5.61}$$

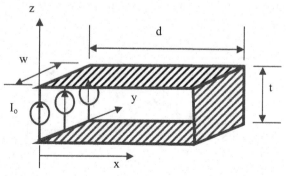

FIGURE 5.13: Lumped element inductor.

The zero order magnetic flux enclosed by the PEC loop connected to the terminals is expressed as

$$\Psi_{m0} = \int\limits_{z=0}^{t} \int\limits_{x=0}^{d} B_0 \, dx dz = \mu H_0 t d = \frac{\mu t d}{w} I_0. \qquad (5.62)$$

With inductance defined as the ratio of magnetic flux linked by the terminal current to the terminal current, i.e., $L = \Psi_m/I$, the DC or zero order inductance is calculated as

$$L_{DC} = L_0 = \frac{\Psi_{m0}}{I_0} = \frac{\mu t d}{w}. \qquad (5.63)$$

There is no zero order voltage drop since a DC current flows in lossless PEC electrodes which leads to a zero order electric field as well, $E_0 = 0$.

5.6.2 First Order Term

Now, we will investigate the first order response of the inductor. Recollect that the sources of first order voltages/currents/charges/fields are their zero order counterparts. The time-varying zero order magnetic field, H_0, is related to the first order electric field through Faraday's law as

$$\oint\limits_{L} \mathbf{E}_1 \cdot \mathbf{dl} = -j\omega \iint\limits_{S} \mu \mathbf{H}_0 \cdot \mathbf{ds}. \qquad (5.64)$$

The zero order magnetic field penetrates the dashed surface indicated in Fig. 5.14. Since H_0 is uniform, the right-hand integral is readily evaluated as

$$-j\omega \iint\limits_{S} \mu \mathbf{H}_0 \cdot \mathbf{ds} = -j\omega \mu H_0 t (d - x). \qquad (5.65)$$

The path for the line integral is fixed to a clockwise direction by the right-hand rule.

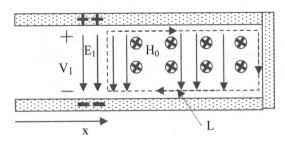

FIGURE 5.14: Integration path to calculate V_1.

The portion of the path that is coincident with the PEC conductor has zero voltage drop so that only the voltage drop along the left end of the path contributes to the integral to give

$$\oint_L \mathbf{E}_1 \cdot \mathbf{dl} = V_1(x). \tag{5.66}$$

When these two previous expressions are equated, the voltage is expressed as

$$V_1(x) = j\omega\mu H_0 t (d - x) = j\omega\mu I_0 t \frac{(d - x)}{w}. \tag{5.67}$$

The voltage increases linearly from the short circuit at the right end of inductor to a peak value at the terminals. The time-varying, zero order magnetic field induces a first order electric field that increases linearly with increasing distance from the right end of the inductor where the electric field is zero at the right end of the inductor adjacent to the PEC. This increase is due to an increasing amount of magnetic flux enclosed by the path L as the distance from the short circuit increases. This behavior is analogous to the behavior of the first order magnetic field in the capacitor.

This behavior is what we would normally expect from an inductor, an induced voltage that is positive at the terminal into which the current flows. Moreover, using the expression for the zero order inductance, the voltage–current relationship at the inductor terminals becomes that used in circuits as

$$V_1(0) = j\omega\mu I_0 t \frac{d}{w} = j\omega L_0 I_0 = j\omega L_{\text{DC}} I_0. \tag{5.68}$$

This quasi-static behavior is the usual low-frequency behavior of an inductor and is based upon the DC inductance. The energy associated with the first order electric field leads to increasingly important capacitive effects as the frequency of operation is increased.

5.6.3   Second Order Term
The first order electric field is expressed as

$$E_1(x) = -j\omega\mu H_0 (d - x) = -j\omega\mu I_0 \frac{(d - x)}{w}. \tag{5.69}$$

In addition, a first order surface charge density is induced by the first order electric field with positive charge on the upper surface and negative on the bottom:

$$\rho_{S1}(x) = D_1(x) = \varepsilon E_1(x)$$
$$= -j\omega\mu\varepsilon H_0 (d - x) = -j\omega\mu\varepsilon I_0 \frac{(d - x)}{w}. \tag{5.70}$$

Let's continue to the second order terms. As seen earlier, the second order terms are excited by the first order terms, i.e., first order voltage and surface charge density, which act as sources for second order currents and magnetic fields. The relationship between first order surface charge density and second order current can be treated similar to the charge–current distribution on a capacitor electrode. This leads to the relationship of

$$\frac{d I_2(x)}{dx} = -j\omega \rho_{s1} w = -j\omega \left(-j\omega \mu \varepsilon \frac{(d-x)}{w} I_0 \right) w$$

$$= -\omega^2 \mu \varepsilon (d-x) I_0. \tag{5.71}$$

A simple integration leads to

$$I_2(x) = - \int_{x'=x}^{d} \omega^2 \mu \varepsilon I_0 (d-x') dx' = -\omega^2 \mu \varepsilon \frac{(d-x)^2}{2} I_0. \tag{5.72}$$

Note that the current at the terminals, $I_2(0)$, has a negative sign, indicating that the second order current is flowing out of the upper terminal. The input current is expressed as

$$I_{0,2}(0) = I_0 + \omega^2 I_2(0) = I_0 \left(1 - \omega^2 \mu \varepsilon \frac{(d-x)^2}{2} \right). \tag{5.73}$$

From these results, the second order input impedance can be written as

$$Z_2 = \frac{V_1(0)}{I_{0,2}(0)} = \frac{j\omega \mu t \frac{d}{w}}{1 - \frac{\omega^2 \mu \varepsilon d^2}{2}} = \frac{j\omega L_0}{1 - \frac{\omega^2 L_0 C_0}{2}} = \frac{1}{\frac{1}{j\omega L_0} + j\omega \frac{C_0}{2}}. \tag{5.74}$$

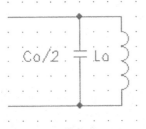

FIGURE 5.15: Second order model for an inductor.

i.e., the parallel combination of L_0 and $C_0/2$, see Fig. 5.15.

In a manner similar to that for the open-circuited electrodes, the short-circuited electrodes can be represented by $L_0 = L_{DC}$ for DC and low frequencies. This quasi-static model is useful for frequencies up to those that approach the resonance of the L_0 and $C_0/2$ elements. This predicts an absolute upper limit for the short-circuited electrodes to function as an inductor of

$$d_{\text{resonance}} = \frac{\sqrt{2}}{2\pi} \lambda \approx 0.225\lambda. \tag{5.75}$$

At this frequency, the input impedance of the short circuit becomes an open circuit.

5.6.4 Higher Order Terms

The second order current is related to the second order surface current density and the second order magnetic field as

$$K_2(x) = H_2(x) = \frac{I_2(x)}{w} = -\omega^2 \mu \varepsilon \frac{(d-x)^2}{2w} I_0. \tag{5.76}$$

The second order magnetic flux enclosed by the second order current is calculated as

$$\Psi_{m2}(x) = -\int_{z=0}^{t} \int_{x'=x}^{d} \mu H_2(x')\, dx'dz$$

$$= -\int_{z=0}^{t} \int_{x'=x}^{d} -\omega^2 \mu^2 \varepsilon \frac{(d-x')^2}{2w} I_0\, dx'dz$$

$$= -\omega^2 \mu^2 \varepsilon \frac{(d-x)^3}{2 \cdot 3w} I_0. \tag{5.77}$$

The negative sign preceding the surface integral indicates that the flux is directed out of the plane of the paper. This second order flux acts as the source of the third order voltage; the procedure is similar to earlier work. The resulting third order voltage is

$$V_3(x) = -\omega^2 \mu^2 \varepsilon \frac{(d-x)^3}{2 \cdot 3w} t I_0. \tag{5.78}$$

The third order impedance is calculated as

$$Z_3 = \frac{V_{1,3}(0)}{I_{0,2}(0)} = \frac{j\omega \mu t \dfrac{d}{w} - \omega^2 \mu^2 \varepsilon \dfrac{d^3}{2 \cdot 3w} t}{1 - \dfrac{\omega^2 \mu \varepsilon d^2}{2}}$$

$$= \frac{j\omega L_0 \left(1 - \omega^2 \dfrac{L_0 C_0}{6}\right)}{1 - \dfrac{\omega^2 L_0 C_0}{2}}. \tag{5.79}$$

As we discovered earlier, though the impedance is unique, the topology of the realization is not. A Foster type II synthesis for this impedance is an inductor of L_0 in parallel with the series connection of an inductor of $L_0/2$ and a capacitor $C_0/3$ as shown in Fig. 5.16.

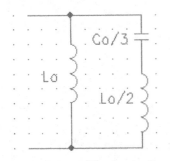

FIGURE 5.16: Third order Foster type II model for a lumped inductor.

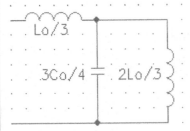

FIGURE 5.17: Third order Cauer type I model for a lumped inductor.

A Cauer type I synthesis results in an inductor of $L_0/3$ in series with the parallel combination of a capacitor of $3C_0/4$ and an inductor of $2L_0/3$ as shown in Fig. 5.17.

The inclusion of the infinite number of terms to account for all orders leads to an expression for the current at the electrode terminals of

$$
I_{\text{Terminals}} = I_0 \left(1 - \frac{\omega^2 \mu \varepsilon d^2}{2!} + \frac{\left(\omega\sqrt{\mu\varepsilon}d\right)^4}{4!} + \frac{\left(\omega\sqrt{\mu\varepsilon}d\right)^6}{6!} - \cdots \right)
$$

$$
= I_0 \cos\left(\omega\sqrt{\mu\varepsilon}d\right) = I_0 \cos(kd) \tag{5.80}
$$

in a manner similar to the voltage of the open-circuited electrodes, the exact wave solution for the electrodes. The terminal voltage is expressed as

$$
V_{\text{Terminals}} = V_1 \left(\omega\sqrt{\mu\varepsilon}d - \frac{\left(\omega\sqrt{\mu\varepsilon}d\right)^3}{3!} + \frac{\left(\omega\sqrt{\mu\varepsilon}d\right)^5}{5!} - \cdots \right)
$$

$$
= V_1 \sin(kd) \tag{5.81}
$$

and the input impedance for shorted electrodes is expressed as

$$
Z_{\text{INSC}} = \frac{V_{\text{Terminals}}}{I_{\text{Terminals}}} = \frac{V_1}{I_0}\tan(kd) = \frac{V_1}{I_0}\tan(\omega\sqrt{\mu\varepsilon}d), \tag{5.82}
$$

the reciprocal of the behavior of open-circuited electrodes.

5.7 RESISTIVE TERMINATIONS

Finally, consider the case of a resistive termination connected between the right edges of the two PEC electrodes as shown at right. With a DC voltage source, V_0, attached to the terminals at the left end of the PEC electrodes, a voltage drop exists between the two electrodes. This in turn sets up a zero order electric field, E_0, between the electrodes that is directed downward. In addition, the voltage drop supports a zero order current, I_0, flowing through the PEC electrodes and the resistive material of the termination. A zero order magnetic field, H_0, directed into the paper is associated with the current flow. These details are shown in Fig. 5.18. The analysis of this configuration will proceed much as in the earlier cases, but with both zero order electric and magnetic fields.

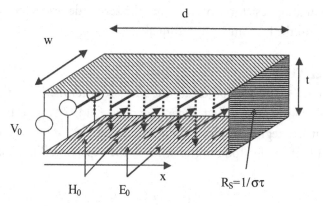

FIGURE 5.18: Lumped element resistor.

5.7.1 Zero Order Term

The zero order voltage source will maintain a voltage drop V_0 across the resistive termination resulting in a zero order electric field directed downward of

$$E_0 = \frac{V_0}{t}.$$ (5.83)

The resistive termination with conductivity σ and thickness τ has a resistance per square of R_S ohms/meter. This leads to a total resistance of

$$R_{\text{Load}} = R_S \frac{t}{w} = \frac{t}{\sigma w \tau}.$$ (5.84)

The zero order current flowing through the resistive termination is calculated with Ohm's law as

$$I_0 = \frac{V_0}{R_{\text{Term}}} = \frac{V_0}{\dfrac{t}{\sigma w \tau}} = \frac{\sigma w \tau V_0}{t}$$ (5.85)

and, as before, a zero order magnetic field directed into the paper is established as

$$H_0 = K_0 = \frac{I_0}{w} = \frac{\sigma \tau V_0}{t}.$$ (5.86)

The zero order resistance of the system is measured at the left edge of the PEC input terminals as

$$R_0 = \frac{V_0}{I_0} = \frac{V_0}{\dfrac{\sigma w \tau V_0}{t}} = \frac{t}{\sigma w \tau} = R_{\text{Load}},$$ (5.87)

the value of the terminating resistor. This becomes the zero order model for the system, a shunt resistor from the top to the bottom input terminal.

5.7.2   First Order Term

There are two zero order sources that create the first order terms. First, there is a zero order electric field, E_0, that generates a first order magnetic field, H_1. The first order current is constant throughout the region between the PEC electrodes, unlike the current that varies linearly from one end of an electrode to the other in the capacitor. The first order current can be expressed as

$$I_1 = \frac{j\omega V_0 w\varepsilon d}{t} = j\omega V_0 C_0. \qquad (5.88)$$

The total of the zero and first order currents is expressed as

$$I_{01} = I_0 + I_1 = I_0 + \frac{j\omega V_0 w\varepsilon d}{t}$$
$$= I_0 + j\omega I_0 R_0 C_0 = I_0 \left(1 + j\omega R_0 C_0\right). \qquad (5.89)$$

The first order voltage due to the zero order magnetic field is constant throughout the region between the electrodes and is expressed as

$$V_1 = j\omega I_0 \frac{\mu d t}{w} = j\omega I_0 L_0. \qquad (5.90)$$

The total of the zero and first order voltages is expressed as

$$V_{01} = V_0 + j\omega I_0 \frac{\mu d t}{w} = V_0 + \frac{j\omega V_0 L_0}{R_0}$$
$$= V_0 \left(1 + \frac{j\omega L_0}{R_0}\right). \qquad (5.91)$$

The quasi-static input impedance (which includes zero and first order terms) is calculated as the ratio of V_{01} and I_{01} as

$$Z_{01} = \frac{V_{01}}{I_{01}} = \frac{V_0 \left(1 + \dfrac{j\omega L_0}{R_0}\right)}{I_0 \left(1 + j\omega R_0 C_0\right)}$$
$$= R_0 \frac{1 + \dfrac{j\omega L_0}{R_0}}{1 + j\omega R_0 C_0}. \qquad (5.92)$$

In general, the quasi-static input impedance varies with frequency.

When $R_0 \ll [L_0/C_0]^{1/2}$, the denominator becomes approximately 1 and the quasi-static input impedance is approximated by only the numerator as

$$Z_{01} = R_0 \frac{1 + \frac{j\omega L_0}{R_0}}{1 + j\omega R_0 C_0} \approx R_0 + j\omega L_0. \qquad (5.93)$$

The series combination of R_0 and L_0 dominates the performance of the element. The capacitance has such a relatively large impedance that is has no significant effect; see Fig. 5.19.

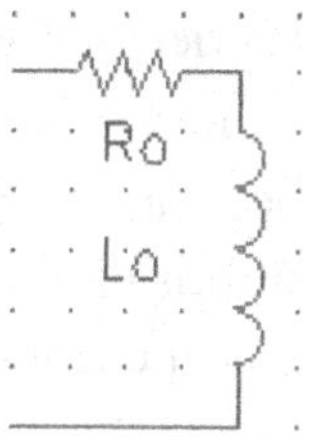

On the other hand, when $R_0 \gg [L_0/C_0]^{1/2}$, the numerator becomes approximately 1 and the quasi-static impedance is approximated by only the denominator as

$$Z_{01} = R_0 \frac{1 + \frac{j\omega L_0}{R_0}}{1 + j\omega R_0 C_0} \approx \frac{R_0}{1 + j\omega R_0 C_0}$$

$$= \frac{1}{\frac{1}{R_0} + j\omega C_0}. \qquad (5.94)$$

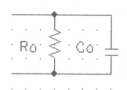

FIGURE 5.19: Lumped resistor model valid for $R_0 \ll [L_0/C_0]^{1/2}$.

The parallel combination of R_0 and C_0 dominate the performance of the element. The inductance has such a relatively small impedance that it has no significant effect as shown in Fig. 5.20.

In the special case where the imaginary parts of the numerator and denominator are equal, i.e.,

$$1 + \frac{j\omega L_0}{R_0} = 1 + j\omega R_0 C_0 \Rightarrow R_0 = \sqrt{\frac{L_0}{C_0}}, \qquad (5.95)$$

the quasi-static input impedance becomes real and equal to the load resistance so that

$$Z_{01} = R_0 = \sqrt{\frac{L_0}{C_0}}. \qquad (5.96)$$

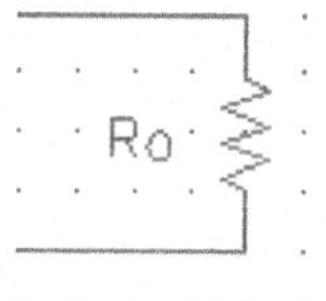

FIGURE 5.20: Lumped resistor model valid for $R_0 \gg [L_0/C_0]^{1/2}$.

This condition is noted as the *matched condition*, i.e., the load resistance is "matched" to the transmission line. Under this condition, the two first order effects produce compensating effects that "cancel" each other. The model for the matched resistor is shown in Fig. 5.21.

The property of the electrodes that describes this behavior is known as the characteristic impedance of the structure. For long, two-electrode structures known as transmission lines, this is called

FIGURE 5.21: Lumped resistor valid for the matched condition, $R_0 = [L_0/C_0]^{1/2}$.

the *characteristic impedance* of the transmission line and is defined as

$$Z_0 = \sqrt{\frac{L_0}{C_0}}. \tag{5.97}$$

Typical instrumentation transmission lines have a characteristic impedance of 50 Ω; cable TV systems have a characteristic impedance of 75 Ω.

These brief glimpses into the electromagnetic behavior of objects that are in the order of a wavelength in size have shown some rather surprising results. But the fun is just beginning. As objects become several wavelengths or more in size, their electromagnetic properties become even more surprising. Waves propagate from point to point, even in space. Why are you waiting? Turn to the next chapter and let's see what other exciting phenomena lurk in electromagnetic fields!

CHAPTER 6

Electromagnetic Waves

6.1   HELMHOLTZ'S WAVE EQUATION

We are now poised to glimpse the excitement experienced by electromagneticians during the late 19th century—propagation of electromagnetic waves. Since this is our first view of electromagnetic waves, we will make several simplifying, though realistic, assumptions. We limit our considerations to lossless, homogeneous media for which $\sigma = 0$ and $\nabla\varepsilon = \nabla\mu = 0$. We will only consider isotropic media with identical characteristics in all directions. Further, let's consider the material to be linear so that we can use superposition of solutions. Moreover, there will be no sources, i.e., $\rho_V = 0$ and $\mathbf{J} = 0$. This last limitation may seem foolish since there can be no fields if there are no sources. However, we can still make an analysis of the characteristic behavior of the system for a unit amplitude signal in much the same way as we examine the transfer function of a network for a unit input signal. Finally, we will consider sinusoidal time dependence and use the phasor form of Maxwell's equations.

The first step is to take the curl of Faraday's law, Eq. (5.6), which gives

$$\nabla \times \nabla \times \mathbf{E} = -\nabla \times j\omega\mathbf{B} = -j\omega\nabla \times \mathbf{B} = -j\omega\mu\nabla \times \mathbf{H} \tag{6.1}$$

in phasor form. The LHS can be rewritten via a vector identity from Appendix A as

$$\nabla \times \nabla \times \mathbf{E} = \nabla(\nabla \cdot \mathbf{E}) - \nabla^2\mathbf{E} = -\nabla^2\mathbf{E} \tag{6.2}$$

where the $\nabla \cdot \mathbf{E} = (\nabla \cdot \mathbf{D})/\varepsilon = 0$ since $\rho_V = 0$. The RHS of Eq. (6.1) can be simplified by substituting the phasor form of Eq. (5.12) to give

$$-j\omega\mu\nabla \times \mathbf{H} = -j\omega\mu\left(\mathbf{J} + j\omega\mathbf{D}\right) = \omega^2\mu\mathbf{D} = \omega^2\mu\varepsilon\mathbf{E} \tag{6.3}$$

where we have used $\mathbf{J} = 0$. Substituting Eqs. (6.2) and (6.3) into Eq. (6.1), we obtain

$$\nabla^2\mathbf{E} + \omega^2\mu\varepsilon\mathbf{E} = 0, \tag{6.4}$$

a single equation involving the electric field intensity only. This form is called the *Helmholtz equation* or the *Helmholtz wave equation* in honor of Heinrich Helmholtz and his pioneering work in establishing the existence of electromagnetic waves. We could have begun with the curl of Ampere's law, Eq. (5.12), to obtain an identical equation in $\mathbf{H}$. The solution of Eq. (6.4) will describe the behavior of both the electric and magnetic fields.

As is our practice, let's solve a simplified version first to get an idea of what to expect. First assume that there is only an x-component of $\mathbf{E}$ so that we have only a scalar equation,

$$\nabla^2 E_X + \omega^2 \mu \varepsilon E_X = 0. \tag{6.5}$$

This form of a field is very common and is known as a *linearly polarized* field since there is only one component of the electric field. Equation (6.5) represents an x-polarized wave. As a further simplification, let's assume that E_X varies only in the z-direction (there's no reason for this choice of direction, it just simplifies the work), i.e., $\partial E_X / \partial x = \partial E_X / \partial y = 0$, so that

$$\nabla^2 E_X + \omega^2 \mu \varepsilon E_X = \frac{\partial^2 E_X}{\partial z^2} + \omega^2 \mu \varepsilon E_X = 0. \tag{6.6}$$

Using the operator solution method for DEQs, we assume the solution will be of the form $e^{\pm pz}$ to obtain

$$\frac{\partial^2 E_X}{\partial z^2} + \omega^2 \mu \varepsilon E_X = \left(p^2 + \omega^2 \mu \varepsilon \right) E_X = 0 \tag{6.7}$$

with solutions of $p = \pm j\omega \sqrt{\mu \varepsilon}$. This leads to the phasor domain solution of

$$E_X(z, j\omega) = E^+(j\omega) e^{-j\omega \sqrt{\mu \varepsilon} z} + E^-(j\omega) e^{j\omega \sqrt{\mu \varepsilon} z}. \tag{6.8}$$

As in circuits, the time domain solution is obtained from the phasor domain solution by multiplying the phasor solution by $e^{j\omega t}$ and retaining only the real part of the result. This approach leads to the time domain solution of

$$\begin{aligned} E_X(z, t) &= Re\left(E^+ e^{-j\omega \sqrt{\mu \varepsilon} z} e^{j\omega t} + E^- e^{j\omega \sqrt{\mu \varepsilon} z} e^{j\omega t} \right) \\ &= |E^+| \cos(\omega t - \omega \sqrt{\mu \varepsilon} z + \phi^+) \\ &\quad + |E^-| \cos(\omega t + \omega \sqrt{\mu \varepsilon} z + \phi^-) \end{aligned} \tag{6.9}$$

where $E_X^+ = |E^+| e^{j\phi^+}$ and $E_X^- = |E^-| e^{j\phi^-}$ are complex constants that may depend upon ω. The magnetic field satisfies the same PDE and has the same form. This result is not the solution to a particular electric field source and a particular configuration of material. Rather, it is the characteristic behavior of a time-harmonic, electromagnetic field in a homogeneous, isotropic region of space. This is very similar to the phasor response of a circuit that describes the characteristic steady-state response of a circuit. Now, let's look at this solution carefully to see why it is called a wave equation.

As expected, the field exhibits sinusoidal time behavior. At a fixed position, say $z = 0$, the field is composed of two sinusoids as

$$E_X(0, t) = |E^+| \cos(\omega t + \phi^+) + |E^-| \cos(\omega t + \phi^-). \tag{6.10}$$

To understand the spatial behavior, consider only the first term of Eq. (6.10). The phase of the cosine term, $\theta^+ = \omega t - \omega\sqrt{\mu\varepsilon}z + \phi^+$, which determines a unique point on the time-varying signal, depends upon both time and space. When $\theta^+ = 0$, this is a peak of the cosine; when $\theta^+ = \pi/2$, it is a zero. A point of nonvarying phase of the cosine exists at locations where the derivative of the phase with respect to time is zero,

$$\frac{d\theta^+}{dt} = \omega - \omega\sqrt{\mu\varepsilon}\frac{dz}{dt} = 0. \tag{6.11}$$

This describes the variation of the z-coordinate of the constant phase point with time. This expression is in terms of the rate of change of position of constant phase point with time, dz/dt, more commonly known as *phase velocity* and expressed as

$$v_p = \frac{dz}{dt} = \frac{1}{\sqrt{\mu\varepsilon}} \quad [\text{m/s}]. \tag{6.12}$$

The constant phase point moves with a positive velocity, i.e., in the positive z-direction. This indicates that a constant phase point, more commonly called a *phase front* (since there are no variations of the wave with respect to x and y), moves or propagates through space with a velocity which is governed by the materials within which it moves. In vacuum, we find that $v_P = 1/[(4\pi \times 10^{-7})(8.854 \times 10^{-12})]^{1/2} \approx 3 \times 10^8$ m/s. We know this as the speed of light in vacuum and denote it as c. This is as expected since light is a high-frequency form of electromagnetic field. We can generalize this expression for materials that have a relative permeability or permittivity greater than 1 as

$$v_p = \frac{c}{\sqrt{\mu_R\varepsilon_R}}. \tag{6.13}$$

The velocity of electromagnetic waves in materials where μ_R, $\varepsilon_R > 1$ (the preponderance of materials) is less than the speed of light in vacuum.

The first term of Equation (6.10) represents a wave propagating in the +z-direction and it is clear that the complex constant E^+ is associated with a positively propagating wave. In a similar way, the second term has an equal, but negative, velocity, directed in the $-z$-direction and represents a negatively propagating wave. The electric field can be composed of a positively and a negatively propagating wave. This decomposition of time-varying fields into two oppositely directed waves is quite convenient and common.

Substituting Eq. (6.12) into the first term of Eq. (6.9), we obtain

$$E_X^+(z, t) = |E^+| \cos\left(\omega t - \frac{\omega}{\sqrt{\mu\varepsilon}}z + \phi^+\right)$$
$$= |E^+| \cos(\omega t - \beta z + \phi^+) \tag{6.14}$$

where β is defined as

$$\beta = \omega\sqrt{\mu\varepsilon} = \frac{\omega}{v_P} \quad [\text{rad/m}], \tag{6.15}$$

the *propagation constant* of the wave with units of radians/meter. For the positively propagating wave of Eq. (6.14), the phase can be expressed as

$$\theta^+ = \omega t - \omega\sqrt{\mu\varepsilon}z + \phi^+ = \omega\left(t - \frac{z}{v_P}\right) + \phi^+$$
$$= \omega(t - t_d) + \phi^+ \tag{6.16}$$

where $t_d = z/v_P$ is a *time delay* in the wave due to the finite phase velocity of the wave. The larger z, the longer the delay since the wave must travel farther.

With these definitions, the phasor form of the field, Eq. (6.8), becomes

$$E_X(z, j\omega) = E^+ e^{-j\beta z} + E^- e^{-j\beta z}. \tag{6.17}$$

The positively propagating wave is described by $e^{-j\beta z}$; the negative-going wave is described by $e^{+j\beta z}$. The exponential term represents phase delays of the wave as it propagates. In general, a wave propagating in the positive coordinate direction will have a negative phase term while the wave propagating in the negative coordinate direction will have a positive sign on the phase term. The phase of both terms is becoming more negative with increasing distance from $z = 0$. These forms are associated with the $e^{j\omega t}$ form of time dependence. Many physics texts use a time dependence of the form $e^{-i\omega t}$ so that they have the opposite signs on the phase terms.

As shown in Eqs. (6.16) and (6.17), the phase of the waves depends only upon the z-coordinate, i.e., the phase is the same at all points on the plane $z = z_o$. Consequently, this wave is known as a *plane wave*. Furthermore, since the amplitude of the field is also constant on this planar surface, it is known as a *uniform plane wave*. As we shall see shortly, this plane can be oriented in other Cartesian directions. In addition, there are *cylindrical and spherical waves*. Though it is impossible for plane waves to exist in nature, they closely approximate many actual situations. In addition, they exhibit the essential features of wave propagation and they are mathematically simpler to analyze. So, we will confine our analysis to plane waves.

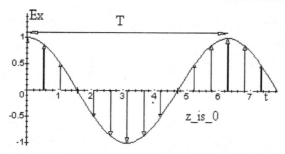

FIGURE 6.1: Temporal variations of the positive-going wave of Eq. (6.14) for $\phi^+ = 0$.

Figure 6.1 represents Eq. (6.14) at $z = 0$ for the special case of $\phi^+ = 0$. Each vector represents the magnitude of the electric field at the location $z = 0$ at various values of time for a wave with radian frequency $\omega = 1$. Sinusoids repeat themselves for phase differences of $\Delta\theta = \pm 2n\pi$ with n an integer. The change of phase due to time variations is expressed as $\Delta\theta^+ = \omega\Delta t$. The time interval between equal phase points is called a period, T, of the signal and is shown in Fig. 6.1. This leads to the well-known expression of

$$T = \frac{2\pi}{\omega} = \frac{1}{f} \quad [\text{s}]. \tag{6.18}$$

The change of phase due to spatial variations is expressed as $\Delta\theta^+ = \beta\Delta z$. The spatial interval between equal phase points is called a *wavelength*, λ, of the propagating wave and is shown in Fig. 6.2. This leads to a comparable expression as

$$\lambda = \frac{2\pi}{\beta} = \frac{v_P}{f} \quad [\text{m}]. \tag{6.19}$$

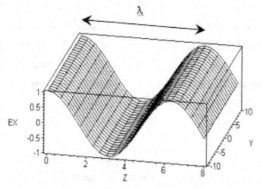

FIGURE 6.2: Spatial variations of the positive-going wave of Eq. (6.14) for $\phi^+ = 0$.

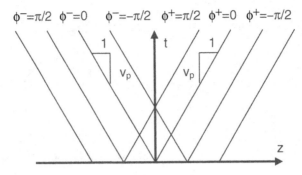

FIGURE 6.3: Space–time trajectories for various phase fronts.

The results for the negative-going portion of the electric field are identical to the temporal variations of Fig. 6.1. As expected, the spatial variations similar to Fig. 6.2 show propagation in the opposite direction.

An alternative view which shows the spatial and temporal behavior on the same graph is shown in Fig. 6.3 where the two axes are time and space. Each instant of time corresponds to a unique spatial location for a particular phase front. The trajectory of all points for the data in Figs. 6.1 and 6.2 is a straight line through the origin with a slope equal to $\pm 1/v_P$; the positively propagating wave has a positive velocity and the negatively propagating wave has a negative velocity. The trajectories for various other phases are shown, also. The intersection of a vertical line with the positive trajectories provides the location for a temporal waveshape at a particular spatial location comparable to Fig. 6.1. The intersection of a horizontal line with the positively sloped trajectories provides the locus for a spatial waveshape for a particular time comparable to Fig. 6.2.

Example 6.1-1. The electric field of a commercial FM signal is approximated as an x-polarized plane wave propagating in the $+z$-direction. The field has an amplitude of 10 µV/m. The frequency of operation is 100 MHz. Obtain a mathematical description for the phasor and time domain forms of this signal. Assume that the wave is propagating in free space and that the phase associated with the wave is zero. The propagation constant is calculated as $\beta = \omega\sqrt{\mu\varepsilon} = 2\pi \times 10^8/3 \times 10^8 = 2.09$ rad/m. The vector direction is $\mathbf{a}_X$; the amplitude is 10^{-5} V/m. The phasor form of the field can be expressed as $\mathbf{E} = \mathbf{a}_X 10^{-5} e^{-j2.09z}$ V/m. The time domain form is $\varepsilon(t) = Re\{\mathbf{a}_X 10^{-5} e^{-j2.09z} e^{j\omega t}\} = 10\mathbf{a}_X \cos[6.28 \times 10^8 t - 2.09z]$ µV/m.

Example 6.1-2. Calculate the wavelength of the wave of Example 6.1-1. $\lambda = 2\pi/\beta = 2\pi/2.09 = 3$ m.

6.2   TEM WAVES AND IMPEDANCE

The positively propagating electric field is accompanied by a corresponding magnetic field that can be readily calculated from the phasor form of Faraday's law, Table 5.2, as

$$\mathbf{H}^+ = \frac{\nabla \times \mathbf{E}^+}{-j\omega\mu} = \frac{\mathbf{a}_Y \dfrac{\partial E_X^+}{\partial z}}{-j\omega\mu} = \frac{-\mathbf{a}_Y j\beta E^+ e^{-j\beta z}}{-j\omega\mu}$$

$$= \left(\frac{\beta}{\omega\mu}\right) \mathbf{a}_Y E^+ e^{-j\beta z} = \left(\frac{\beta}{\omega\mu}\right) \mathbf{a}_Y E_X^+$$

$$= \frac{\mathbf{a}_Y E_X^+}{\eta^+} \tag{6.20}$$

where $\eta^+ = E^+/H^+ = \omega\mu/\beta = [\mu/\varepsilon]^{1/2}.\eta^+$ has units of ohms and is defined as the *wave impedance*. There are no terminals present so there is not a measured impedance as in a circuit. However, since $\eta^+ = E_X^+/H_Y^+$ has units of ohms (volts per meter/amperes per meter) it is called an impedance. More generally, wave impedance is defined as the ratio of an electric field intensity component to a magnetic field intensity component. In free space, $\eta = 120\pi = 377\ \Omega$. In addition, the cross product, $\mathbf{E} \times \mathbf{H}$, points in the direction the wave is propagating. For the negatively propagating portion of the wave, we have

$$\mathbf{H}^- = \frac{\nabla \times \mathbf{E}^-}{-j\omega\mu} = \frac{\mathbf{a}_Y \dfrac{\partial E_X^-}{\partial z}}{-j\omega\mu} = \frac{\mathbf{a}_Y j\beta E^- e^{j\beta z}}{-j\omega\mu}$$

$$= -\left(\frac{\beta}{\omega\mu}\right) \mathbf{a}_Y E^- e^{j\beta z} = -\left(\frac{\beta}{\omega\mu}\right) \mathbf{a}_Y E_X^-$$

$$= \frac{\mathbf{a}_Y E_X^-}{\eta^-} \tag{6.21}$$

with $\eta^- = E_X^-/H_Y^- = -[\mu/\varepsilon]^{1/2}$. The minus sign is due to the magnetic field directed in the $-y$-direction; the negative sign is NOT due to a negative impedance in the material. Since we only consider isotropic materials in this textbook, the wave propagates the same in all directions. The only differences in sign are due to the mathematical representations.

The magnetic field is perpendicular to the electric field as shown by Eqs. (6.20) and (6.21). Moreover, they are both perpendicular to the direction of propagation. In this situation, we have transverse electric and magnetic (TEM) fields to the direction of propagation. This is typical of all electromagnetic waves—planar, cylindrical, and spherical—in space where there are no charges present.

As shown by Eqs. (6.20) and (6.21), the electric and magnetic fields have the same propagation constant and are in phase with each other. This is typical of all electromagnetic waves in lossless, charge-free regions as well.

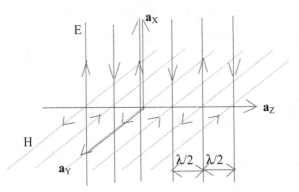

FIGURE 6.4: Interlinked fields of plane wave propagating in the $+z$-direction.

The magnetic fields associated with TEM plane waves are linked to the corresponding electric fields as discussed earlier. For the positive-going wave, the electric field lines are oriented in the $+x$-direction while the magnetic field lines are in the $+y$-direction, see Fig. 6.4. Of course the magnetic fields must close on themselves. In addition, in the absence of charge, the electric fields close upon themselves, too. Both sets of field lines must extend to infinity because the phase is constant over the entire plane. One-half wavelength away, the field lines are oppositely directed as we observed in Fig. 6.2. At infinity, these oppositely directed field lines join together to form a closed, albeit infinite, loop. The x-directed electric field lines enclose y-directed magnetic field lines that in turn enclose additional electric field lines that in turn enclose additional magnetic field lines and so on ad infinitum. This structure of interlinked, closed field lines repeats itself every wavelength. In addition, the field lines are moving with velocity v_P in the $+z$-direction. Of course, there is a similar structure of field lines associated with the negatively propagating wave as well.

Let's review what we know about linearly polarized plane waves in lossless, homogeneous, isotropic, linear, source-free regions. The electric and magnetic fields are perpendicular to each other and they are perpendicular or transverse to the direction of propagation. Their cross product, $\mathbf{E} \times \mathbf{H}$, points in the direction of propagation. Constant phase fronts of these fields occupy planes and propagate together with a constant velocity that is determined by the material in which they propagate as $\pm 1/[\mu\varepsilon]^{1/2}$; the sign is determined by the direction of propagation. The ratio of the field phasors is constant and determined by the material properties according to $E_X^+/H_Y^+ = \eta^+ = [\mu/\varepsilon]^{1/2}$ and $E_X^-/H_Y^- = \eta^- = -[\mu/\varepsilon]^{1/2}$. The wave repeats itself every $\lambda = v_P/f$ in the direction of propagation.

Example 6.2-1. Calculate the accompanying phasor form of the magnetic field intensity for the signal of Example 6.1-1. The tried and true method is to use

Faraday's law which gives $\mathbf{H} = \nabla \times \mathbf{E}/(-j\omega\mu) = -j2.09(10^{-5})\mathbf{a}_Y e^{-j2.09z}/(-j2\pi \times 10^8(4\pi \times 10^{-7}) = \mathbf{a}_Y(0.0265)e^{-j2.09z}$ μA/m. An alternative method is to use the properties of plane waves. The direction of $\mathbf{H}$ must be such that $\mathbf{E} \times \mathbf{H}$ points in the $\mathbf{a}_Z$-direction; this direction must be $\mathbf{a}_Y$. The magnetic field in phasor form is calculated as the phasor electric field divided by η^+; $H_Y = 10^{-5}/377 = 0.0265$ μA/m. The phase variation will be the same as the electric field. This leads to $\mathbf{H} = \mathbf{a}_Y(0.0265)e^{-j2.09z}$ μA/m as we found before.

Example 6.2-2. Calculate the phasor magnetic field associated with a signal similar to that of Example 6.1-1 but propagating in the $-z$-direction. The magnetic field will have the same magnitude, but will have an opposite sign since $\mathbf{E} \times \mathbf{H}$ must point in the $-z$-direction. The phase term will be opposite as well due to the opposite direction of propagation. Consequently, the magnetic field intensity is given by $\mathbf{H} = -\mathbf{a}_Y(0.0265)e^{j2.09z}$ A/m.

6.3   SOURCES OF TEM WAVES

So far we have assumed that the plane waves exist and are propagating in a source-free region. But, how can waves be initiated? Consider an infinite current sheet located in the $z = 0$ plane on which a sinusoidal surface current has a phasor form $\mathbf{K} = -K_o\mathbf{a}_X$, see Fig. 6.5.

Due to symmetry, we expect the magnetic fields on opposite sides of the current sheet to be equal, but oppositely directed. Ampere's law (or boundary conditions) allows us to calculate these values as we did in Section 3.15 to obtain

$$|\mathbf{H}|_{\text{SURFACE}} = \frac{K_o}{2}. \tag{6.22}$$

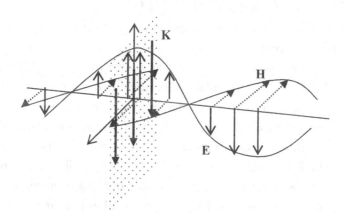

FIGURE 6.5: A current sheet as the source of electromagnetic waves.

We have seen that two waves are possible—positively propagating and negatively propagating. But the physical realities of the problem suggest that the waves will only travel away from the source which means

$$\mathbf{H} = -\mathbf{a}_Y \frac{K_o}{2} e^{j\beta z}, \quad z < 0$$

$$= \mathbf{a}_Y \frac{K_o}{2} e^{-j\beta z}, \quad z > 0. \tag{6.23}$$

Ampere's law is always valid for calculating the electric field from a known magnetic field. But, for the case of plane waves, we can use their properties that we identified in the last section to calculate the electric field. $\mathbf{E}$ will have the same exponential form as $\mathbf{H}$ and it will be polarized so that $\mathbf{E} \times \mathbf{H}$ is in the $+z$-direction for $z > 0$ and the $-z$-direction for $z < 0$. The phasor value of $\mathbf{E}$ is calculated by multiplying the phasor value of $\mathbf{H}$ by η. From these steps we obtain

$$\mathbf{E} = \mathbf{a}_X \frac{\eta K_o}{2} e^{j\beta z}, \quad z < 0$$

$$= \mathbf{a}_X \frac{\eta K_o}{2} e^{-j\beta z}, \quad z > 0. \tag{6.24}$$

Notice that it takes an infinite plane of current to generate the plane wave. Nothing of lesser dimensions can do so, though good approximations in limited regions can be achieved with finite current distributions. Moreover, a true plane wave carries an infinite amount of power; to establish it would take all of the power in the universe! Obviously, this is impractical. Since plane waves are mathematically simple and display the fundamental principles of wave propagation, they serve a useful purpose as model from which we can learn and against which we can compare other forms of waves.

6.4   WAVES IN ARBITRARY DIRECTIONS

Our wave calculations have been relatively simple when the direction of propagation coincides with a coordinate axis. But how can we handle waves in other directions, such as the wave moving at an angle θ with respect to the z-axis as shown in Fig. 6.6. The direction of the wave propagation is denoted as z'. Since plane waves are TEM, the electric and magnetic fields must lie in the plane perpendicular to z' denoted by AA'. Any electric field in this plane can be composed of a linear combination of $E_\perp$ and $E_\parallel$ (perpendicular and parallel to the plane of incidence formed by the direction of propagation and the normal to the $z = 0$ plane) with the corresponding magnetic fields oriented such that $\mathbf{E} \times \mathbf{H}$ is in the direction of propagation. For convenience, let's define the electric field as $E_\perp$ totally in the $\mathbf{a}_Y$-direction, i.e., directed out of the plane of the paper. The corresponding magnetic field has an amplitude given by $E_\perp/\eta$, lies in the $x-z$ plane, and is directed toward the lower right-hand corner of the page, i.e., with

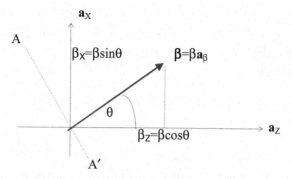

FIGURE 6.6: A plane wave in an arbitrary direction.

components in the $-\mathbf{a}_X$ and $+\mathbf{a}_Z$ directions. The wave nature is exhibited by the imaginary term in the exponential as $e^{-j\beta z}$. Though all these details are correct, it is often more useful to represent the fields in the $x-y-z$ coordinates instead of the primed coordinates.

It is clear that z' has both x and z components; their relative amplitudes depending upon the angle θ. The components can be determined from the geometry of the situation as

$$z' = z\cos\theta + x\sin\theta, \tag{6.25}$$

a simple coordinate transformation. This enables us to write the electric field as

$$\mathbf{E}_\perp = \mathbf{a}_Y E_{o\perp} e^{-j\beta(z\cos\theta + x\sin\theta)}$$
$$= \mathbf{a}_Y E_{o\perp} e^{-j\beta z\cos\theta} e^{-j\beta x\sin\theta}. \tag{6.26}$$

This form discloses that the wave is represented as propagating in two directions—partially in the z-direction, partially in the x-direction. The corresponding magnetic field is given by

$$\mathbf{H}_\perp = (-\mathbf{a}_X\cos\theta + \mathbf{a}_Z\sin\theta)\frac{E_{o\perp}}{\eta} e^{-j\beta z\cos\theta} e^{-j\beta x\sin\theta} \tag{6.27}$$

where the properties of TEM plane waves have been used to compute the components.

In a similar way, the parallel components can be determined as

$$\mathbf{H}_\| = \mathbf{a}_Y \frac{E_{o\|}}{\eta} e^{-j\beta z\cos\theta} e^{-j\beta x\sin\theta} \tag{6.28}$$

and

$$\mathbf{E}_\| = (\mathbf{a}_X\cos\theta - \mathbf{a}_Z\sin\theta)E_{o\|} e^{-j\beta z\cos\theta} e^{-j\beta x\sin\theta}. \tag{6.29}$$

The propagation at an angle with respect to the z-axis can be generalized to an arbitrary direction. For the cases shown in Fig. 6.6, we generalize the scalar propagation constant to a vector, β, pointing in the direction of propagation with a magnitude of $\omega\sqrt{\mu\varepsilon}$. Then, we can regroup the exponential terms as

$$\beta z \cos\theta + \beta x \sin\theta = \beta \cos\theta z + \beta \sin\theta z$$
$$= \beta_Z z + \beta_X x \qquad (6.30)$$

where β_Z and β_X are the z and x components of the β vector. Using this case as an example, we can generalize further to propagation in three dimensions where β is expressed as

$$\beta = \mathbf{a}_\beta \omega\sqrt{\mu\varepsilon} = \beta_X \mathbf{a}_X + \beta_Y \mathbf{a}_Y + \beta_Z \mathbf{a}_Z$$
$$= \omega\sqrt{\mu\varepsilon}(\cos\gamma_X \mathbf{a}_X + \cos\gamma_Y \mathbf{a}_Y + \cos\gamma_Z \mathbf{a}_Z). \qquad (6.31)$$

$\gamma_i = \mathbf{a}_i \cdot \mathbf{a}_\beta$ is the angle between the ith unit vector and the propagation vector β. When the dot product of the propagation vector, β, and the position vector, $\mathbf{r} = x\mathbf{a}_X + y\mathbf{a}_Y + z\mathbf{a}_Z$, is formed, the result is

$$\beta \cdot \mathbf{r} = \beta_X x + \beta_Y y + \beta_Z z. \qquad (6.32)$$

This is the form of the exponent in the phase term. For the geometry of Fig. 6.6, $\gamma_X = \pi/2 - \theta$, $\gamma_Y = \pi/2$, and $\gamma_Z = \theta$, so that $\beta_X = \sin\theta$, $\beta_Y = 0$, and $\beta_Z = \cos\theta$. This method enables representation of waves propagating in any direction. As with a wave propagating along the z-axis, the sign of each component determines the direction along that axis—a negative sign indicates that the wave is moving in the positive-axis direction, and a positive sign indicates motion in the negative-axis direction. The general representation of a wave moving in the direction β is given as

$$\mathbf{E} = \mathbf{a}_E E_o e^{-j\beta \cdot \mathbf{r}}. \qquad (6.33)$$

In addition, since this is a TEM plane wave, the electric and magnetic fields must be perpendicular to the direction of propagation and to each other so that

$$\mathbf{a}_E \cdot \mathbf{a}_H = \mathbf{a}_E \cdot \beta = \mathbf{a}_H \cdot \beta = 0. \qquad (6.34)$$

Finally, the cross product of the electric and magnetic fields must point in the direction of propagation as

$$\mathbf{a}_E \times \mathbf{a}_H = \mathbf{a}_\beta. \qquad (6.35)$$

With these relationships, we can represent any TEM wave propagating in any direction. This representation is useful when the arbitrary plane waves are impingent on planar boundaries.

Example 6.4-1. A TEM plane wave is moving from the origin into the $x, y, z > 0$ quadrant with $\beta_X = \beta_Y = \beta_Z$. The frequency is 300 MHz and the material is air. The electric field has $E_X = E_Y = 1$ V/m. Obtain expressions for the electric and magnetic fields associated with this wave. The propagation constant has a magnitude given by $\beta = \omega\sqrt{\mu\varepsilon} = 2\pi$ so that the propagation vector is given as $\beta = 2\pi(\mathbf{a}_X + \mathbf{a}_Y + \mathbf{a}_Z)/\sqrt{3}$ and the exponential form becomes $e^{-j\beta\cdot\mathbf{r}} = e^{-j2\pi(\mathbf{a}_X+\mathbf{a}_Y+\mathbf{a}_Z)\cdot(x\mathbf{a}_X+y\mathbf{a}_Y+z\mathbf{a}_Z)/\sqrt{3}} = e^{-j(2\pi/\sqrt{3})(x+y+z)}$. The vector portion of the electric field is expressed as $\mathbf{E}_o = (\mathbf{a}_X + \mathbf{a}_Y + E_Z\mathbf{a}_Z)$. Note that there may be a z-component of the electric field. Since the electric field and the propagation vector must be perpendicular, we find that $\mathbf{E}_o \cdot \beta = [(\mathbf{a}_X + \mathbf{a}_Y + E_Z\mathbf{a}_Z) \cdot 2\pi(\mathbf{a}_X + \mathbf{a}_Y + \mathbf{a}_Z)/\sqrt{3} = (2\pi/\sqrt{3})(1 + 1 + E_Z) = 0$. Therefore, $E_Z = -2$ and the phasor form of the electric field is $\mathbf{E} = \mathbf{a}_X + \mathbf{a}_Y + \mathbf{a}_Z e^{-j2\pi/\sqrt{3}(x+y+z)}$ V/m. The magnetic field has a magnitude $|\mathbf{H}| = |\mathbf{E}|/\eta = \sqrt{6}/377 = 6.5$ mA/m; the direction is governed by $\mathbf{a}_E \times \mathbf{a}_H = \mathbf{a}_\beta$ which is equivalent to $\mathbf{a}_H = \mathbf{a}_\beta \times \mathbf{a}_E = (\mathbf{a}_X + \mathbf{a}_Y + \mathbf{a}_Z)/\sqrt{3} \times (\mathbf{a}_X + \mathbf{a}_Y - 2\mathbf{a}_Z)/\sqrt{6} = (-\mathbf{a}_X + \mathbf{a}_Y)/\sqrt{2}$. These results lead to $\mathbf{H} = 6.5/\sqrt{2}(-\mathbf{a}_X + \mathbf{a}_Y)e^{-j(2\pi/\sqrt{3})(x+y+z)}$ mA/m. Alternatively, we can calculate $\mathbf{H}$ from the differential form of Ampere's law as $\mathbf{H} = \nabla \times \mathbf{E}/(-j\omega\mu)$ to obtain identical results.

6.5   WAVES IN LOSSY MATERIAL

We have excluded most materials and a number of significant effects by limiting our study to lossless materials only. For example, instead of analyzing the effects of losses in copper wires, we consider them to be PECs with insignificant effects on most solutions. However, the effects of losses become more pronounced for high frequency signals. We must consider the effects of losses on wave propagation.

This is easily accomplished by allowing $\sigma \neq 0$. Following the methods of Section 6.1, first we take the curl of Faraday's law to obtain

$$\nabla^2\mathbf{E} = j\omega\mu\nabla \times \mathbf{H} = j\omega\mu(\mathbf{J} + j\omega\mathbf{D})$$
$$= j\omega\mu(\sigma + j\omega\varepsilon)\mathbf{E} \tag{6.36}$$

where we have used $\mathbf{J} = \sigma\mathbf{E}$ and $\mathbf{D} = \varepsilon\mathbf{E}$. For a linear, x-polarized $\mathbf{E}$, this becomes

$$\frac{\partial^2 E_X}{\partial z^2} - j\omega\mu(\sigma + j\omega\varepsilon)E_X = \frac{\partial^2 E_X}{\partial z^2} - \gamma^2 E_X = 0 \tag{6.37}$$

where γ is known as the *complex propagation constant*. Solutions to this equation are of the form $e^{\pm\gamma z}$. From our previous solutions for positive-going waves of the form $e^{-j\omega t}$, we will "guess" that the $e^{-\gamma z}$ form will represent waves in the +z-direction. Consequently, the positive-going solution becomes

$$\mathbf{E} = \mathbf{a}_X E^+ e^{-\gamma z} \tag{6.38}$$

where

$$\gamma = \sqrt{j\omega\mu(\sigma + j\omega\varepsilon)} = j\omega\sqrt{\mu\varepsilon}\sqrt{1 + \frac{\sigma}{j\omega\varepsilon}} \tag{6.39}$$

in its most general form. Numeric results for this form are easily obtained, but we must make simplifications in order to make some general observations. The condition of $\sigma/\omega\varepsilon \ll 1$ is commonly called the *low-loss case*; such materials are known as *good dielectrics*. Lossless materials for which $\sigma = 0$ are often called *perfect dielectrics*. Materials with $\sigma/\omega\varepsilon \gg 1$ is the *high-loss* (or just *lossy*) case; such a material is better known as a *good conductor*. Of course when $\sigma = \infty$ we call the material a PEC.

For the *low-loss case*, $\sigma/\omega\varepsilon \ll 1$, the right-most term of Eq. (6.39) is approximated by the binomial theorem as $[1 + \sigma/j\omega\varepsilon]^{1/2} \approx 1 + \sigma/j2\omega\varepsilon = 1 - j\sigma/2\omega\varepsilon$. This leads to the electric field solution of

$$\mathbf{E} = \mathbf{a}_X E^+ e^{-j\omega\sqrt{\mu\varepsilon}\left(1 - \frac{j\sigma}{2\omega\varepsilon}\right)z} = \mathbf{a}_X E^+ e^{-\left(\frac{j\omega\sqrt{\mu\varepsilon}j\sigma}{2\omega\varepsilon} + j\omega\sqrt{\mu\varepsilon}\right)z}$$
$$= \mathbf{a}_X E^+ e^{-\sqrt{\frac{\mu}{\varepsilon}}\frac{\sigma}{2}z - j\omega\sqrt{\mu\varepsilon}z} = \mathbf{a}_X E^+ e^{-\alpha z - j\beta z} = \mathbf{a}_X E^+ e^{-\gamma z} \tag{6.40}$$

where $\alpha = \mathrm{Re}(\gamma)$ and $\beta = \mathrm{Im}(\gamma)$ and

$$\gamma = \sqrt{\frac{\mu}{\varepsilon}}\frac{\sigma}{2} + j\omega\sqrt{\mu\varepsilon}. \tag{6.41}$$

As in the lossless case, the $-j\beta z$ term in the exponential provides the phase shift with distance and gives the field its wave nature. This solution form verifies our intuition that small losses should not alter the wave nature of the field. The $-\alpha z$ term in the exponential is purely real and doesn't contribute to the wave nature of the field at all. Instead, it causes the wave to decrease in amplitude as it propagates. α is known as the *attenuation constant* and has units of nepers/meter (Np/m). Nepers is a dimensionless unit (just as radians is dimensionless) introduced by Bell Laboratory employees in the 1930s in honor of the Swiss mathematician, Napier, and his work with natural logarithms. The amplitude of the field is reduced as it propagates because part of its power is absorbed by the lossy material. In fact, this loss of power by the wave heats the material. Note that as the conductivity of the medium approaches zero and the material becomes lossless, the wave propagates unattenuated as expected.

In the *high-loss case*, $\sigma/\omega\varepsilon \gg 1$, and the right-most term of Eq. (6.39) can be approximated as $[1 + \sigma/j\omega\varepsilon]^{1/2} \approx [\sigma/j\omega\varepsilon]^{1/2}$. This leads to the complex propagation constant of the form

$$\gamma = j\omega\sqrt{\mu\varepsilon}\sqrt{1 + \frac{\sigma}{j\omega\varepsilon}} \approx j\omega\sqrt{\mu\varepsilon\frac{\sigma}{j\omega\varepsilon}} = \sqrt{j\omega\mu\sigma}$$

$$= \sqrt{\omega\mu\sigma}\sqrt{j} = \sqrt{j\omega\mu\sigma}\sqrt{e^{j\frac{\pi}{2}}} = \sqrt{\omega\mu\sigma}\,e^{j\frac{\pi}{4}}$$

$$= \left(\frac{1+j}{\sqrt{2}}\right)\sqrt{\omega\mu\sigma} = \sqrt{\frac{\omega\mu\sigma}{2}} + j\sqrt{\frac{\omega\mu\sigma}{2}}. \tag{6.42}$$

For this case the attenuation and propagation constants are equal and they increase with increasing frequency. Since this occurs in good conductors, i.e., copper wires, it is of great importance; we will examine this phenomenon later.

The wave impedance is affected by losses as well. The magnetic field accompanying the electric field in Eq. (6.38) can be calculated via Faraday's law as

$$\mathbf{H}^+ = \frac{\nabla \times \mathbf{E}^+}{-j\omega\mu} = \mathbf{a}_Y\frac{E_X e^{-\gamma z}}{\left(\dfrac{j\omega\mu}{\gamma}\right)} = \mathbf{a}_Y\frac{E_X e^{-\gamma z}}{\eta^+}. \tag{6.43}$$

The general form of the wave impedance for positively propagating waves in materials with loss is expressed as

$$\eta^+ = \frac{j\omega\mu}{\gamma} = \sqrt{\frac{j\omega\mu}{\sigma + j\omega\varepsilon}}. \tag{6.44}$$

For *low-loss materials*, the wave impedance becomes

$$\eta^+ = \frac{j\omega\mu}{\gamma} = \frac{j\omega\mu}{j\omega\sqrt{\mu\varepsilon}\sqrt{1 + \dfrac{\sigma}{j\omega\varepsilon}}} \approx \sqrt{\frac{\mu}{\varepsilon}}\left(1 - \frac{j\sigma}{2\omega\varepsilon}\right). \tag{6.45}$$

The last term is often negligible and the wave impedance is virtually unaltered from the lossless case. In the *high-loss case*, the wave impedance becomes

$$\eta^+ = \frac{j\omega\mu}{\gamma} = \frac{j\omega\mu}{j\omega\sqrt{\mu\varepsilon}\sqrt{1 + \dfrac{\sigma}{j\omega\varepsilon}}} \approx \sqrt{\frac{\mu}{\varepsilon}\frac{j\omega\varepsilon}{\sigma}}$$

$$= \sqrt{\frac{\omega\mu}{\sigma}}\,e^{j\frac{\pi}{4}} = (1 + j)\sqrt{\frac{\omega\mu}{2\sigma}}. \tag{6.46}$$

The wave impedance for the high-loss case shows that the magnetic field has a $-45°$ phase shift relative to the electric field. This is due to the stored magnetic energy associated with current

TABLE 6.1: Plane Wave Parameters.

General form	$\gamma = \sqrt{j\omega\mu(\sigma + j\omega\varepsilon)}$	$\eta^+ = \sqrt{\dfrac{j\omega\mu}{\sigma + j\omega\varepsilon}}$
Perfect dielectric, $\sigma = 0$	$\gamma = j\omega\sqrt{\mu\varepsilon}$	$\eta^+ = \sqrt{\dfrac{\mu}{\varepsilon}}$
Low-loss dielectric, $\sigma/\omega\varepsilon \ll 1$	$\gamma = \sqrt{\dfrac{\mu}{\varepsilon}}\dfrac{\sigma}{2} + j\omega\sqrt{\mu\varepsilon}$	$\eta^+ \approx \sqrt{\dfrac{\mu}{\varepsilon}}$
Good conductor, $\sigma/\omega\varepsilon \gg 1$		$\eta^+ = (1+j)\sqrt{\dfrac{\omega\mu}{2\sigma}}$

flow. Moreover, the wave impedance increases with increasing frequency. This is especially important when we consider fields in good conductors such as copper.

The details for wave propagation in three types of materials—perfect dielectrics, low-loss dielectrics, and good conductors—are shown in Table 6.1.

We have considered waves propagating in the $+z$-direction only. The behavior will be the same for propagation in the opposite direction (or in other directions); attenuation and phase shift will occur as the wave propagates. The signs of these two terms will be opposite for opposite direction of propagation.

Even metals cease to act as good conductors at high enough frequencies for which $\sigma/\omega\varepsilon \ll 1$. Consider copper which has $\sigma = 5.8 \times 10^7$ S/m and $\varepsilon = \varepsilon_o = 8.854 \times 10^{-12}$ F/m. At a frequency of $f \approx 1 \times 10^{18}$ Hz, the copper ceases to act as a good conductor. This is somewhat above the frequency of light. At these frequencies, electromagnetic waves are no longer reflected as in a mirror.

Losses can also occur due to the materials with complex permittivity or permeability. As can be seen from Eq. (5.57), the complex propagation constant, γ, is purely imaginary for the lossless case when μ and ε are real and $\sigma = 0$. On the other hand, γ becomes complex and the material is lossy when μ or ε are complex. This is usually due to some molecular, atomic, or electronic resonance that causes the material to absorb power such as the heating of food in a microwave oven. But, we will have to defer discussion of this behavior till a later time.

Example 6.5-1. Calculate the propagation constant and wave impedance for a plane wave in material with μ_o, $\varepsilon_R = 8$, and $\sigma = 10^{-5}$ S/m for operation 1 MHz. From Table 6.1, the propagation constant is given by

$$\gamma = \sqrt{j\omega\mu(\sigma + j\omega\varepsilon)}$$

$$= \sqrt{j2\pi \times 10^6(4\pi \times 10^{-7})(10^{-5} + j2\pi \times 10^6 \times 8 \times 8.854 \times 10^{-12})}$$

$$= (0.6659 + j59.29)10^{-3}/\text{m}.$$

The wave impedance is calculated as

$$\eta^+ = \sqrt{\frac{j\omega\mu}{\sigma + j\omega\varepsilon}} = \sqrt{\frac{j2\pi \times 10^6(4\pi \times 10^{-7})}{10^{-5} + j2\pi \times 10^6 \times 8 \times 8.854 \times 10^{-12}}}$$

$$= 133.2 + j1.5 \ \Omega.$$

Example 6.5-2. A linearly polarized wave operates at a frequency of 1 MHz. The wave is propagating in a lossy material characterized by $\gamma = 0.00067 + j0.059$ m^{-1} and $\eta = 133.2 + j1.5 \ \Omega$. Calculate the material properties μ, ε, and σ. Using the general representations for γ, Eq. (6.89), and η, Eq. (6.90), we can calculate $j\omega\mu = [\gamma\eta]^{1/2}$ and $\sigma + j\omega\varepsilon = [\gamma/\eta]^{1/2}$. These calculations yield $\mu = \mu_o$, $\sigma = 10^{-5}$ S/m, and $\varepsilon = 8\varepsilon_o$. Note that these calculations are accurate to two decimal places only.

6.6 GOOD CONDUCTORS AND SKIN DEPTH

The behavior of time-varying electromagnetic fields in the vicinity of good conductors is of importance to all electrical engineers. Contrary to our common assumption that copper wires can be treated as PECs, in this section they are better classified as good conductors. Consider a wave propagating in the +z-direction into an infinite half-plane of good conductor as shown in Fig. 6.7. A linearly polarized electric field is expressed as

$$\mathbf{E} = \mathbf{a}_\chi E_o e^{-\gamma z} = \mathbf{a}_\chi E_o e^{-\sqrt{\frac{\omega\mu\sigma}{2}}z}e^{-j\sqrt{\frac{\omega\mu\sigma}{2}}z} \qquad (6.47)$$

where E_o is the amplitude of the field at the surface of the good conductor and the good conductor approximation for the propagation constant of Table 6.1 is used. The first exponential term represents attenuation with propagation; the second, phase shift which gives the field a wave nature. The rate of attenuation is frequently described in the same manner by which the duration of transient signals is described—at what depth has the field strength decreased to e^{-1} of its value at the surface? The depth at which this occurs is defined as

$$\delta = \sqrt{\frac{2}{\omega\mu\sigma}} = \sqrt{\frac{1}{\pi f \mu\sigma}} \qquad (6.48)$$

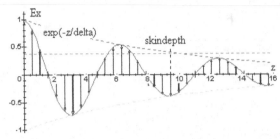

FIGURE 6.7: Plane wave propagation into a good conductor.

and is called the *skin depth*. The field penetrates the conductor to infinity, but for values of $z \gg \delta$ it has a negligibly small value. The greater the frequency, the less the skin depth. At one wavelength of penetration $z = 2\pi$ and the amplitude of the field has been reduced to a value of $e^{-2\pi} \approx 0.002$, its amplitude at the conductor surface. For copper, Eq. (6.48) becomes

$$\delta_{CU} = \sqrt{\frac{1}{\pi f \mu \sigma}} = \sqrt{\frac{1}{\pi f (4\pi \times 10^{-7})(5.8 \times 10^6)}} = \frac{0.067}{\sqrt{f}}. \qquad (6.49)$$

The skin depth of copper is given as a function of frequency in Table 6.2.

TABLE 6.2: Skin Depth for Copper at Several Frequencies.

Frequency, f (Hz)	100	10 k	100 M	10 G
Skin depth, δ (m)	0.0067	6.7×10^{-4}	6.7×10^{-6}	6.7×10^{-7}

Equation (6.47) can be rewritten in terms of δ as

$$\mathbf{E} = \mathbf{a}_X E_o e^{-\gamma z} = \mathbf{a}_X E_o e^{-\frac{z}{\delta}} e^{-j\frac{z}{\delta}} \qquad (6.50)$$

where $\alpha = \beta = 1/\delta$.

The current density and the magnetic field exhibit the same attenuation as well since they are proportional to the electric field. All three fields have significant values only near the surface of the good conductor. The wave impedance is expressed in terms of δ as

$$\eta^+ = \frac{j\omega\mu}{\gamma} = \sqrt{\frac{\omega\mu}{\sigma}} e^{j\frac{\pi}{4}} = (1+j)\sqrt{\frac{\omega\mu}{2\sigma}} = \frac{1+j}{\sigma\delta}. \qquad (6.51)$$

The wave impedance is complex with positive real and imaginary parts indicating that the current flow in the conductor has both a resistive and an inductive component. The resistive part, $1/\sigma\delta$, is often called *wave resistance* or *resistance/square* and has units of ohms. The **H** field lags the **E** field by $\pi/4$ due to the inductive component of the wave impedance.

The mathematics clearly indicates a high attenuation of the wave as it propagates into the good conductor, but what is the physical explanation of this effect? The time-varying electric field establishes a current, aligned with the electric field, which in turn establishes a perpendicular magnetic field that has a phase delay of $\pi/2$. This time-varying magnetic field induces an electric field which is perpendicular and phase delayed by $\pi/2$ from the magnetic field. This induced electric field is aligned with but of opposite sign to the original electric field. Hence, the total electric field is reduced. The induced electric field sets up an additional current that in turn establishes a further magnetic field and so on ad infinitum. This process is initiated by the original transient wave when it begins propagating into the conductive region. The steady-state result is an attenuated wave propagating into the conducting half-plane. The electromagnetic field is "shielded" from the interior of the conducting material by the currents that the fields establish. This effect is present in all materials for which the conductivity is nonzero; however, the results are significant in good conductors only.

PECs are characterized by $\sigma = \infty$ for which $\delta \to 0$. Therefore, no time-varying electromagnetic fields penetrate the material. The current also flows in a vanishingly thin layer on the surface of the PEC. In this case, it is described as a surface current density rather than a volume current density. Moreover, since the magnetic field within the PEC is zero, it satisfies the boundary condition

$$\mathbf{a}_N \times \mathbf{H}_1 = \mathbf{K}. \tag{6.52}$$

The time-varying magnetic field at the surface of a PEC establishes a surface current on the surface equal in magnitude to the magnetic field intensity adjacent to the surface.

The phenomenon of skin depth has a significant effect on high-frequency circuits, which is the subject of the next section.

Example 6.6-1. The electric field propagating into a semi-infinite block of copper is expressed by $\mathbf{E} = \mathbf{a}_X e^{-\frac{z}{\delta}} e^{-j\frac{z}{\delta}}$. Obtain an expression for the magnetic field. From Faraday's law or from the properties of TEM plane waves, the magnetic field is given by

$$\mathbf{H} = \mathbf{a}_Y \frac{1+j}{j\omega\mu\delta} e^{-\frac{z}{\delta}} e^{-j\frac{z}{\delta}} = \mathbf{a}_Y \frac{1-j}{\omega\mu\delta} e^{-\frac{z}{\delta}} e^{-j\frac{z}{\delta}}.$$

6.7 SKIN EFFECT IN CIRCUITS

The tendency for current to flow only near the surface of conductors greatly affects high-frequency circuit performance. The effects for a planar surface as discussed in the previous section must be modified somewhat for cylindrical geometry. Rather than solving the wave equation in cylindrical coordinates, let's use an iterative method to obtain the fields of a long

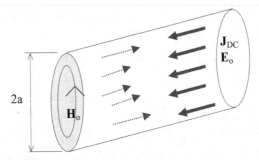

FIGURE 6.8: Time-varying fields in acylindrical conductor.

cylindrical wire. We will assume a uniform current distribution in the wire as we had for DC currents; this is valid when the fields are not changing. But for time-varying fields, Maxwell's equations require that additional "correction" terms be added to describe time-varying currents. The process is similar to the development of higher order terms for lumped element circuits in Chapter 5.

Consider a cylindrical conductor oriented along the z-axis as shown in Fig. 6.8. The wire has a radius of $\rho = a$ meters and a conductivity of σ S/m. The first approximation of the current is uniform and directed along the z-axis (shown as the thicker, wider vectors in Fig. 6.8) as

$$\mathbf{J} = \mathbf{a}_Z J_{DC}. \tag{6.53}$$

The resulting angularly symmetric magnetic field intensity can be calculated by

$$\oint_L \mathbf{H} \cdot \mathbf{Dl} = \int_{\phi=0}^{2\pi} H_\phi \rho d\phi = 2\pi \rho H_\phi = \iint_S \mathbf{J} \cdot \mathbf{ds}$$

$$= \int_{\phi=0}^{2\pi} \int_{\rho=0}^{a} J_{DC} \rho d\rho d\phi = J_{DC}\pi\rho^2, \tag{6.54}$$

which gives

$$H_{\phi o} = \frac{\rho J_{DC}}{2}. \tag{6.55}$$

The induced electric field due to this changing magnetic field is calculated from Faraday's law as

$$\nabla \times \mathbf{E} = -\mathbf{a}_\phi \frac{\partial E_{Z o}}{\partial \rho} = -\mathbf{a}_\phi j\omega\mu H_{\phi o} = -j\mathbf{a}_\phi \omega\mu \frac{\rho J_{DC}}{2}. \tag{6.56}$$

By noting that there is no magnetic field at the origin and, hence, no induced electric field there either, we can integrate the second and fourth terms to obtain this correction of the electric field as

$$E_{Z_0} = \int_{\rho'=0}^{\rho} \frac{j\omega\mu J_{DC}}{2} \rho' d\rho' = j\omega\mu J_{DC} \frac{\rho^2}{2^2}. \tag{6.57}$$

This leads to the first correction term for the current density as

$$J_{DC1} = j\omega\mu\sigma \frac{\rho^2}{2^2} J_{DC} = -\left(\sqrt{\frac{-j\omega\mu\sigma}{2}} \frac{\sqrt{2}}{2} \rho\right)^2 J_{DC}$$

$$= -\left(\frac{\left(\frac{\sqrt{-j2}}{\delta}\rho\right)^2}{4}\right) \left(\frac{1}{1!}\right)^2 J_{DC}. \tag{6.58}$$

The unusual form of Eq. (6.58) will prove useful later. When the process of Eqs. (6.54) through (6.57) is applied to J_{DC1}, a second correction term to the current density is obtained as

$$J_{DC2} = \frac{(-j\omega\mu\sigma)^2 \rho^4}{2^2 4^2} J_{DC} = \left(\frac{\left(\frac{\sqrt{-j2}}{\delta}\rho\right)^2}{4}\right)^2 \left(\frac{1}{2!}\right)^2 J_{DC}. \tag{6.59}$$

The process is continued on to give the nth correction term as

$$J_{DCn} = (-1)^n \left(\frac{z^2}{4}\right)^n \left(\frac{1}{n!}\right)^2 J_{DC} \tag{6.60}$$

where $z = (-j2)^{1/2}(\rho/\delta)$. The total current is the sum of the original approximation and all the correction terms as

$$J_Z = J_{DC} \sum_{n=0}^{\infty} (-1)^n \left(\frac{z^2}{4}\right)^n \left(\frac{1}{n!}\right)^2. \tag{6.61}$$

This form may not be familiar, but it is well known to mathematicians as the series form of the zeroth order Bessel function of the first kind, $J_o(z)$, as

$$J_Z = J_o(z) = J_{DC} \sum_{n=0}^{\infty} (-1)^n \left(\frac{z^2}{4}\right)^n \left(\frac{1}{n!}\right)^2. \tag{6.62}$$

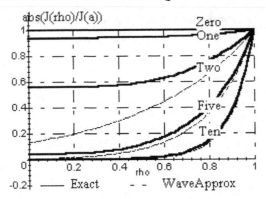

FIGURE 6.9: Current distribution in a cylindrical copper wire.

Note that the mathematicians' symbol for the Bessel function, $J_0(z)$, should not be confused with current densities J_Z and J_{DC}. This is the exact form for the current density in a cylindrical conducting wire; the electric and magnetic fields have the same form. Though Bessel function calculations are readily available in MAPLE, MATLAB, and MATHCAD, they cannot be performed as easily in analytic form as the exponential calculations of the field for planar surfaces of Eq. (6.50). Since skin depths are so small for frequencies above the audio range, as shown in Table 6.2, even wires as small as a millimeter are large compared to skin depth.

When the wire diameter is much larger than skin depth, i.e., $a/\delta \gg 1$, constant radii at the surface of the wire are so large compared to a skin depth that the geometry can be approximated as planar. This suggests that the effects of skin depth in cylindrical wires can be accurately approximated by the simpler, plane-wave form. A comparison of these two methods of calculation is shown in Fig. 6.9 for a unit amplitude current density at the surface of a copper wire. Exact calculations with the Bessel function are shown for $a/\delta = 0, 1, 2, 5$, and 10; approximate plane wave calculations are shown for $a/\delta = 2, 5$, and 10. This shows that nearly identical results are obtained for $a/\delta \geq 10$. The plane wave approximation gives accurate results for 2 MHz and higher with wires of 1 mm diameter. For larger wires, accurate results are obtained for even lower frequencies. We will use this approximation throughout this text. Since the current only "penetrates" a small depth into the wire, our previous calculations of resistance that assumed uniform current density are not valid for AC operations. An alternate approach is to calculate the resistance/square surface area and combine series and parallel squares to obtain the total resistance. To simplify the calculations, we will use the planar model, see Fig. 6.10.

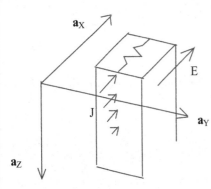

FIGURE 6.10: Resistance/square of the planar surface.

At the surface, voltage drop across the square is calculated by the usual line integral to give $\Delta V \approx |E_S|\Delta l = |J_S|\Delta l/\sigma$ in the direction of the electric field. $\mathbf{E}_S$ and $\mathbf{J}_S$ are the electric field and current density, respectively, at the surface. Though the greatest contribution to the total current is from the current density near the surface, all of the current that flows through the block of conductive material is included. The total current is obtained by the usual surface integral calculation

$$\Delta I = \iint_S \mathbf{J} \cdot \mathbf{ds} = \Delta l \int_{z=0}^{\infty} J_S e^{-\gamma z} dz = \Delta l \frac{J_S}{\gamma}. \tag{6.63}$$

The impedance of the square, the ratio of voltage drop to current flow, is given by

$$Z_w = R_w + jX_w = \frac{\Delta V}{\Delta I} = \frac{\dfrac{J_o \Delta l}{\sigma}}{\dfrac{J_o \Delta l}{\gamma}} = \frac{\gamma}{\sigma} = \frac{1+j}{\sigma \delta}. \tag{6.64}$$

The square has loss associated with the conductivity represented by the real part or the impedance. In addition, it has an inductive component called *internal inductance* (see Example 3.15-3) associated with the stored magnetic energy *within* the conductor due to current flow in it. The impedance/square represents a circuit interpretation of the fields in the conductor. It is identical to the wave impedance of Eq. (6.51), a wave interpretation of the fields. Though both the field and the circuit approaches give the same results, the circuit interpretation is more easily used for current flow in wires. Furthermore, the internal inductance is often negligibly small compared to magnetic energy stored *outside* the conductor, the *external inductance*, and is neglected.

The calculation of the resistance of a wire to AC current flow is simplified by another consideration. Consider the resistance of a square of material of conductivity σ and thickness δ with uniform current density, see Fig. 6.11. The length and width of the square are Δl. The resistance of the square is given by

$$R_w = \frac{L}{\sigma A} = \frac{\Delta l}{\sigma \delta \Delta l} = \frac{1}{\sigma \delta}. \tag{6.65}$$

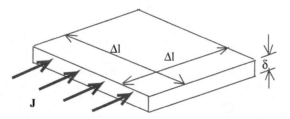

FIGURE 6.11: Resistance of a square conductor with a uniformly distributed current.

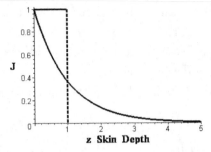

FIGURE 6.12: Amplitude of actual current density and its uniformly distributed equivalent.

This is identical to the real part of the wave impedance and to the exact value of the resistance/square. This suggests an alternative interpretation as a resistor in which all of the current density is uniformly distributed throughout the skin depth with no current flow within the rest of the conductor. The real part of the total resistance (due to the real part of the current) flow is the same in both cases. Figure 6.12 shows the relationship between the amplitudes of the actual and the uniform model for the current density.

This provides another view of the reason for the name of skin depth attached to the thickness of the conductor in which the current flows. The AC resistance of a conductor of any cross section can be calculated using this concept as long as the skin depth is much less than the smallest cross-sectional dimension of the conductor, $\delta/a \ll 1$.

Consider the conductor shown in Fig. 6.13(a); the length is L, the perimeter is P, and the conductivity is σ. The cross-sectional area can be calculated exactly, but since δ is so small, an accurate approximation is obtained by "unwrapping" the skin depth of material from the surface into a flat, sheet of thickness δ as shown in Fig. 6.13(b). From this model, the resistance can be readily calculated as

$$R = \frac{L}{\sigma A} = \frac{L}{\sigma \delta P} = R_w \frac{\frac{L}{\Delta l}}{\frac{P}{\Delta l}} = R_w \frac{N_{\text{SERIES}}}{N_{\text{PARALLEL}}}. \quad (6.66)$$

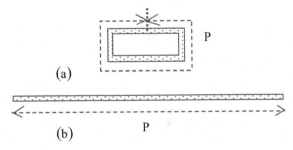

FIGURE 6.13: Two views of skin depth.

The resistance can be calculated analytically using the length and perimeter or graphically by dividing the surface of the conductor into series and parallel squares.

It is obvious that the greater the perimeter, the smaller the resistance due to an increased cross-sectional area for current flow. This strategy is most simply implemented by making the conductor as a thin sheet, but it is not as easily manufactured or as conveniently used as cylindrical wires. Unfortunately, a cylindrical wire minimizes the perimeter for a given cross-sectional area. An alternative strategy is to use stranded wire as shown in Fig. 6.14(a) where N smaller, insulated strands of cylindrical wires are twisted together to form a conductor. When $\delta/a \ll 1$ for each of the strands, the effective perimeter of the stranded wire is N times the circumference of each individual wire. An alternative view of a stranded wire is shown in Fig. 6.14(b).

(a)

(b)

Another strategy to reduce the resistance of a copper wire is to plate it with a thin layer of silver; the greater conductivity of silver will reduce the resistance. The layer of silver needed is only about one skin depth thick so the added cost of the silver is kept to a minimum.

FIGURE 6.14: Stranded wires: (a) cross-section of seven-stranded wire; (b) stranded wire.*

It is obvious that the smaller the diameter of the wire, the greater the resistance. In coaxial transmission lines, the inner conductor contributes most to the resistance. Stranding and silver plating of the center conductor (but not the outer) is often effective enough in reducing the resistance.

Not only is the AC resistance of cylindrical copper wires greater than the DC resistance, it also increases with increasing frequency as shown by Eq. (6.65) as

$$R = \frac{L}{\sigma A_{\mathrm{EFF}}} = \frac{L}{\sigma \delta 2\pi a} = \frac{L}{\sigma \sqrt{\frac{2}{\omega\mu\sigma}} 2\pi a}$$

$$= \frac{L}{2a}\sqrt{\frac{\mu f}{\pi\sigma}}. \tag{6.67}$$

Perhaps more informative is a plot of $R_{\mathrm{AC}}/R_{\mathrm{DC}}$ with respect to frequency. Frequencies for which $\delta/a \gg 1$, the ratio is nearly unity as the skin effect has little effect. For frequencies for which $\delta/a \gg 1$, the ratio increases with $\sqrt{f}$ as

$$\frac{R_{\mathrm{AC}}}{R_{\mathrm{DC}}} = \frac{\dfrac{L}{\sigma \delta 2\pi a}}{\dfrac{L}{\sigma \pi a^2}} = \frac{a}{2\delta} = \frac{a}{2}\sqrt{\pi f \mu \sigma}. \tag{6.68}$$

*Contributed by Ryan Vande Water, EE'96, Rose-Hulman.

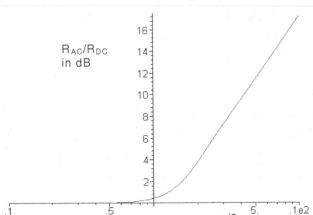

FIGURE 6.15: R_{AC}/R_{DC} versus frequency.

The numeric results for the exact resistances shown in Fig. 6.15 indicate these two asymptotic responses. Note that since the ratio varies as the $\sqrt{f}$, the slope of the high-frequency asymptote is 10 dB/decade.

Example 6.7-1. Calculate the skin depth for silicon steel which has $\mu \approx 0.005$ H/m and $\sigma \approx 10^7$ S/m for operation at 60 Hz. From Eq. (6.49), skin depth is calculated as $\delta = \sqrt{\frac{1}{\pi f \mu \sigma}} = \sqrt{\frac{1}{\pi 60(0.005) \times 10^7}} = 0.33$ mm. The conductivity is not nearly as large as that of copper, but the very large permeability causes the skin depth to be very small.

Example 6.7-2. A copper wire of 1 mm diameter is to be silver plated to reduce its resistance/meter for operation at 1 MHz. What thickness of silver should be used? What is the resistance/meter with this layer? The conductivity of silver is 6.17×10^7 S/m. Since we approximate the current flow as confined to a skin depth. This seems an appropriate thickness and is calculated from Eq. (6.49) as $\delta = \sqrt{\frac{1}{\pi f \mu \sigma}} = \sqrt{\frac{1}{\pi 10^6 (4\pi \times 10^{-7}) 6.17 \times 10^7}} = 64.1$ μm. This leads to a resistance/meter (calculated from Eq. (6.67)) of

$$R/m = \frac{1}{\sigma \delta 2\pi a} = \frac{1}{6.17 \times 10^7 (64.1 \times 10^{-6}) 2\pi (0.0005)}$$
$$= 80.5 \text{ m}\Omega/\text{m}.$$

A copper wire without the silver plating would have a resistance/meter of

$$R/m = \frac{1}{\sigma \delta 2\pi a} = \frac{\sqrt{10^6}}{5.8 \times 10^7 (0.067) 2\pi (0.0005)}$$
$$= 81.9 \text{ m}\Omega/\text{m}.$$

This slight improvement in resistance/meter significantly reduces the attenuation properties of the wire as part of a transmission line.

Example 6.7-3. Calculate the resistance/meter of the copper wire of Example 6.7-2 which uses a seven-stranded wire of the same diameter as the single wire. Each strand will have a diameter of approximately 1/3 the diameter of the solid wire. Therefore, the resistance/meter of each strand will be 81.9(3) = 0.246 mΩ/m. But there are seven strands in parallel so the overall resistance of the stranded wire is 35.1 mΩ/m compared to 81.9 mΩ/m for the single wire.

6.8   POWER FLOW AND POYNTING'S VECTOR

Propagating electromagnetic waves emanate from time-varying sources and move through materials. These moving waves carry energy from sources to distant points, such as the electromagnetic energy from the sun reaching the earth or the power of a radio transmitter reaching a listener. The mechanism for this power flow is described in this section.

Working in the time domain, we take the dot product of $\boldsymbol{\mathcal{E}}$ with the differential form of Ampere's law as

$$\boldsymbol{\mathcal{E}} \cdot \nabla \times \boldsymbol{\mathcal{H}} = \boldsymbol{\mathcal{E}} \cdot \boldsymbol{\mathcal{J}} + \boldsymbol{\mathcal{E}} \cdot \frac{\partial \boldsymbol{\mathcal{D}}}{\partial t} \tag{6.69}$$

and the dot product of $\boldsymbol{\mathcal{H}}$ with the differential form of Faraday's law as

$$\boldsymbol{\mathcal{H}} \cdot \nabla \times \boldsymbol{\mathcal{E}} = -\boldsymbol{\mathcal{H}} \cdot \frac{\partial \boldsymbol{\mathcal{B}}}{\partial t}. \tag{6.70}$$

The difference between these two equations gives

$$\boldsymbol{\mathcal{H}} \cdot \nabla \times \boldsymbol{\mathcal{E}} - \boldsymbol{\mathcal{E}} \cdot \nabla \times \boldsymbol{\mathcal{H}} = -\boldsymbol{\mathcal{E}} \cdot \boldsymbol{\mathcal{J}} - \boldsymbol{\mathcal{E}} \cdot \frac{\partial \boldsymbol{\mathcal{D}}}{\partial t} - \boldsymbol{\mathcal{H}} \cdot \frac{\partial \boldsymbol{\mathcal{B}}}{\partial t}. \tag{6.71}$$

The LHS can be simplified with the vector identity

$$\boldsymbol{\mathcal{H}} \cdot \nabla \times \boldsymbol{\mathcal{E}} - \boldsymbol{\mathcal{E}} \cdot \nabla \times \boldsymbol{\mathcal{H}} = \nabla \cdot (\boldsymbol{\mathcal{E}} \times \boldsymbol{\mathcal{H}}). \tag{6.72}$$

The two terms on the RHS with derivatives can be simplified by

$$\boldsymbol{\mathcal{E}} \cdot \frac{\partial \boldsymbol{\mathcal{D}}}{\partial t} = \varepsilon \boldsymbol{\mathcal{E}} \cdot \frac{\partial \boldsymbol{\mathcal{E}}}{\partial t} = \frac{\varepsilon}{2} \frac{\partial (\boldsymbol{\mathcal{E}} \cdot \boldsymbol{\mathcal{E}})}{\partial t} = \frac{\partial \left(\frac{\varepsilon |\boldsymbol{\mathcal{E}}|^2}{2} \right)}{\partial t}$$
$$= \frac{\partial w_e}{\partial t}. \tag{6.73}$$

and

$$\mathcal{H} \cdot \frac{\partial \mathcal{B}}{\partial t} = \mu \mathcal{H} \cdot \frac{\partial \mathcal{H}}{\partial t} = \frac{\mu}{2} \frac{\partial (\mathcal{H} \cdot \mathcal{H})}{\partial t} = \frac{\partial \left(\frac{\mu |\mathcal{H}|^2}{2} \right)}{\partial t}$$

$$= \frac{\partial w_m}{\partial t} \tag{6.74}$$

where in earlier chapters we defined w_e and w_m as the electric and magnetic energy density, respectively. In addition, the current density term can be divided into two parts, an active component due to a source current density, $\mathcal{J}_{\text{SOURCE}}$, and a passive component due to conduction current, $\sigma \mathcal{E}$. With these modifications, Eq. (6.71) becomes

$$\nabla \cdot (\mathcal{E} \times \mathcal{H})$$
$$= -\mathcal{E} \cdot \mathcal{J}_{\text{SOURCE}} - \mathcal{E} \cdot (\sigma \mathcal{E}) - \frac{\partial (w_e + w_m)}{\partial t}. \tag{6.75}$$

Integration of this equation throughout an arbitrary volume gives

$$\iiint_V \nabla \cdot (\mathcal{E} \times \mathcal{H}) dv = \oiint_S (\mathcal{E} \times \mathcal{H}) \cdot \mathbf{ds}$$
$$= - \iiint_V \left(\mathcal{E} \cdot \mathcal{J}_{\text{SOURCE}} + \sigma |\mathcal{E}|^2 + \frac{\partial (w_e + w_m)}{\partial t} \right) dv \tag{6.76}$$

where the divergence theorem was used on the LHS to convert the volume integral to a closed surface integral. A closer look at the surface integral form of the LHS of Eq. (6.76) reveals that the vector quantity $\mathcal{E} \times \mathcal{H}$ has units of $(V/m)(A/m) = w/m^2$ which represents a power flux density. Moreover, it points perpendicularly to both $\mathcal{E}$ and $\mathcal{H}$, the direction of wave propagation as we have discovered earlier. Therefore, the closed integral over the surface S can be interpreted as the power out of the region V. Since it is represented in terms of the fields $\mathcal{E}$ and $\mathcal{H}$, this power must be the power contained within the fields. The power flux density has been named *Poynting's vector* in honor of English physicist John H. Poynting who brought this remarkable property to light as

$$\mathcal{P} = \mathcal{E} \times \mathcal{H} \, [w/m^2]. \tag{6.77}$$

Nothing has been added to Maxwell's equations; this is an inherent property of electromagnetic fields. In keeping with properties of the divergence of a vector field, the divergence of Poynting's vector, $\nabla \cdot \mathcal{P}$, is zero everywhere except where there are volume-distributed sources. (The integral of this form is always valid, but sources of a point, line, or surface form are mathematically undefined for the divergence operator without some special mathematical definitions.)

The power emanating from the region ∀ is expressed as

$$\oint_S (\mathcal{E} \times \mathcal{H}) \cdot \mathbf{ds} = \text{Power emanating from the region } \forall. \qquad (6.78)$$

The first term in the volume integral of the RHS of Eq. (6.76), $-\mathcal{E} \cdot \mathcal{J}_{\text{SOURCE}}$, is similar to the expression for power dissipation in a resistor,

$$\Delta P_{\text{DISS}} = \Delta V \Delta I = \mathbf{E} \cdot \mathbf{J} \Delta a \Delta l \qquad (6.79)$$

where $\mathbf{E} \cdot \mathbf{J}$ is the volume power density within the conductive material. This represents the power dissipated within a resistor. The negative of this quantity represents the volume power density supplied at a point. For a resistor, the power supplied is negative as it only dissipates power. On the other hand, in a source, the current flow is opposite to the direction of the electric field and it supplies power. Consequently, $-\mathcal{E} \cdot \mathcal{J}_{\text{SOURCE}}$ represents the power density supplied by any sources within the region; the integral represents the total power supplied by sources within the region as

$$\iiint_\forall -\mathcal{E} \cdot \mathcal{J}_{\text{SOURCE}} dv = \text{Power from sources within the region } \forall. \qquad (6.80)$$

The second term within the volume integral, $\sigma |\mathcal{E}|^2$, can be interpreted from Eq. (1.175) as well. It is the power density dissipated at a point due to conduction losses in the material. When integrated throughout the volume, it represents the total ohmic loss, i.e., power dissipation, within the volume ∀,

$$\iiint_\forall \sigma |\mathcal{E}|^2 dv = \text{Power dissipated within the region } \forall. \qquad (6.81)$$

The third term within the volume integral on the RHS of Eq. (6.76), $\partial(w_e + w_m)/\partial t$, is due to the time rate of change of energy density at each point with units of power. This represents the rate at which energy is being added to the energy density stored within the fields. $\partial w_e/\partial t$ represents the rate at which the electric energy density storage is changing; $\partial w_m/\partial t$ represents the magnetic energy density counterpart. The minus sign describes the negative of energy density changes. The integral represents the summation of these energy changes throughout the entire region ∀.

$$\iiint_\forall \frac{\partial(w_e + w_m)}{\partial t} dv = \text{Rate of energy increase within the region } \forall. \qquad (6.82)$$

Putting together the interpretations provided by Eqs. (6.78), (6.79), (6.80), and (6.81), we obtain

$$\underbrace{\oint_S (\boldsymbol{\mathcal{E}} \times \boldsymbol{\mathcal{H}}) \cdot \mathbf{ds}}_{\substack{\text{Power emanating from} \\ \text{the region } \forall \text{ through surface S}}} = \underbrace{-\iiint_\forall \boldsymbol{\mathcal{E}} \cdot \boldsymbol{\mathcal{J}}_{\text{SOURCE}} dv}_{\substack{\text{Power generated by sources} \\ \text{within the region } \forall}}$$

$$\underbrace{-\iiint_\forall \sigma |\boldsymbol{\mathcal{E}}|^2 dv}_{\substack{\text{Power dissipated} \\ \text{within the region } \forall}} \underbrace{-\iiint_\forall \frac{\partial(w_e + w_m)}{\partial t} dv}_{\substack{\text{Rate of energy increase} \\ \text{within the region } \forall}} .$$

(6.83)

The electromagnetic power emanating from within a region is equal to the power generated within the region minus the power dissipated within the region and minus the rate of stored energy increase within the region. This is nothing but a conservation of energy statement that we could obtain by good understanding of the physical processes and sound reasoning. But the good news is that this property is inherent in Maxwell's equations, also, another strong argument that these are a correct representation of electromagnetic fields in nature!

The LHS of Eq. (6.83) describes the power flow across a closed surface in terms of those power generation and absorption mechanisms within the enclosed region. Poynting's vector describes a power flux density that correctly describes the power crossing a closed surface. This is such a plausible representation for the flow of power that it is tempting to apply Poynting's vector to calculate the power flow through any surface, closed or not. For example, Poynting's vector could be integrated over the surface of an imaginary "window pane" in space to find the power flow through it. Though without a rigorous mathematical basis for nonclosed surfaces, such calculations are routinely made and usually yield correct results.

The results of Eq. (6.83) are valid for all time-varying fields. It is appropriate for DC fields that are related to each other, e.g., the electric and magnetic fields associated with the voltage drop and current flow in a resistor. However, DC electric and magnetic fields that exist independently of each other do not emanate power from a closed region, e.g., the combined fields of a permanent magnet and a charged capacitor.

Often we are more interested in time-averaged instead of instantaneous power flow. The usual rules from circuits still apply so that the time-averaged Poynting's vector is expressed as

$$P_{\text{AVE}} = \frac{1}{T} \int_{t'=t_o}^{t_o+T} \boldsymbol{\mathcal{P}}(t') dt'.$$

(6.84)

For sinusoidal signals that are in-phase, the average power is usually the peak instantaneous power divided by 2. In general, we could compute the time-average. But, it is more common practice to put Eq. (6.83) in phasor form so that we can use standard phasor calculation of average power. Recall that in circuits the phasor form of average power is $P_{AVE} = Re\{VI^*/2\}$ which suggests that the average Poynting's vector is defined as

$$P_{AVE} = \frac{Re(\mathbf{E} \times \mathbf{H}^*)}{2}. \tag{6.85}$$

Since the other terms in Eq. (6.83) are all vector dot products, they can be expressed in a similar fashion. The phasor form can be developed as we did with the time-domain form starting with Eqs. (6.69) and (6.70) where $\mathbf{H}^*$ is used in the phasor form. This requires that the conjugate of Ampere's law is used as well. From these steps we obtain

$$\nabla \cdot (\mathbf{E} \times \mathbf{H}^*) = -\mathbf{E} \cdot \mathbf{J}^*_{SOURCE} - \mathbf{E} \cdot (\sigma \mathbf{E}^*)$$
$$- j\omega \left(\mu|\mathbf{H}|^2 - \varepsilon|\mathbf{E}|^2 \right) \tag{6.86}$$

as the phasor form of Eq. (6.75). Integrating throughout volume V, dividing by 2, and using the divergence theorem, we obtain

$$\underbrace{\oiint_S \frac{\mathbf{E} \times \mathbf{H}^*}{2} \cdot \mathbf{ds}}_{\substack{\text{Power emanating from} \\ \text{the region V through the surface S}}} = \underbrace{-\iiint_\forall \frac{\mathbf{E} \cdot \mathbf{J}^*_{SOURCE}}{2} dv}_{\substack{\text{Power generated by sources} \\ \text{within the region V}}}$$

$$\underbrace{-\iiint_\forall \frac{\sigma|\mathbf{E}|^2}{2} dv}_{\substack{\text{Power dissipated} \\ \text{within the region V}}} \underbrace{- j2\omega \iiint_\forall \frac{\mu|\mathbf{H}|^2 - \varepsilon|\mathbf{E}|^2}{4} dv}_{\substack{\text{Rate of energy increase} \\ \text{within the region V}}} \tag{6.87}$$

as the phasor domain counterpart of Eq. (6.83). Phasor power has both a real and imaginary part; the real part is related to the time-averaged power; the imaginary part is related to the reactive power. The LHS is the complex power emanating from the surface S. Similarly, the RHS is the complex power generated, dissipated, and stored within the region $\forall$. A factor of 2 is included in the denominator of all the terms in order to put into the phasor form for average power when peak values for the electric and magnetic fields are used. If expressed as RMS quantities, the factor is unnecessary.

While the first two terms on the RHS are easily understood, consider the last term. It represents the rate of change of stored energy (recall that $j\omega$ in the phasor domain is equivalent to the time derivative in the time domain). An additional factor of 2 is inserted into the numerator as stored energy changes from electrical to magnetic form at a rate of twice the source frequency. Consequently, an additional factor of 2 must be inserted into the denominator as well. The first factor of the integrand, $\mu \left| \mathbf{H} \right|^2$, represents stored magnetic energy; when the term is divided by 2, this represents the peak value of magnetic energy stored in the volume; the magnetic energy varies sinusoidally from 0 to this peak value. This means that the additional factor of 2 in the denominator gives the average magnetic energy stored. The second term, $\varepsilon \left| \mathbf{E} \right|^2$, can be interpreted similarly as electric stored energy. For real values of μ and ε, the third term is imaginary and represents time-averaged reactive power. For cases where the stored energies are equal, there is no reactive power. This is the distributed circuit equivalent of a resonant circuit. In some situations, this happens only at discrete frequencies; in other cases, such as a plane wave, it is valid for all frequencies.

The average power generated or dissipated is often measurable. By using Poynting's vector, we can readily compare the theoretical values with measured values, a very useful connection.

Example 6.8-1. Calculate the complex Poynting's vector for the plane wave of Example 6.1-1. Poynting's vector is readily calculated as $\mathbf{P} = \mathbf{E} \times \mathbf{H}^*/2 = (\mathbf{a}_x 10^{-5} e^{-j2.09z}) \times (\mathbf{a}_y 2.65 \times 10^{-8} e^{-j2.09z})^*/2 = \mathbf{a}_z 1.33 \times 10^{-13}$ w/m^2. Since the plane wave doesn't change in amplitude, Poynting's vector is constant. Moreover, there is only real, average power in the wave, no reactive power.

Example 6.8-2. The electric field far from an omni-directional antenna system in lossless, free space is expressed as $\mathbf{E} = \mathbf{a}_\theta 200 e^{-jkr}/r$ V/m. Calculate the power of the transmitter. The power can be expressed in terms of the complex Poynting's vector for which the magnetic field is required. $\mathbf{H} = \mathbf{a}_\phi 200 e^{-jkr}/377r$ A/m which gives $\mathbf{P} = \mathbf{E} \times \mathbf{H}^*/2 = [\mathbf{a}_\theta 200 e^{-jkr}/r] \times [\mathbf{a}_\phi 200 e^{-jkr}/377r]^*/2 = \mathbf{a}_r (200)^2/754r^2$ w/m^2. The total power emanating from a sphere of radius r is given by

$$P_{\text{AVE}} = \oiint_S \mathbf{P} \cdot \mathbf{ds} = \int_{\phi=0}^{2\pi} \int_{\theta=0}^{\pi} \frac{(200)^2}{754r^2} \mathbf{a}_r \cdot \mathbf{a}_r r^2 \sin\theta \, d\theta \, d\phi$$

$$= 666.7 \text{ w.}$$

Alternatively, since Poynting's vector is uniform over the surface of the sphere, the total power radiated is simply Poynting's vector multiplied by the area of the sphere, 4[greek symbol for PI]r^2. to get the same result.

Example 6.8-3. Calculate both sides of Eq. (6.87) for a plane wave propagating in the lossy material of Example 6.5-1 through a cube with surfaces located at x, y, $z = 0$ and 1. Assume that the wave is linearly polarized in the x-direction with a phasor amplitude of 1 in the $z = 0$ plane, it is propagating in the z-direction, and it is operating at a frequency of 1 MHz. From Example 6.5-1, the propagation constant is $\gamma = 0.00067 + j0.059$ m^{-1} and the wave impedance is $\eta = 133.2 + j1.5$ Ω which are based upon μ_o, $\varepsilon_R = 8$, and $\sigma = 10^{-5}$ S/m. From these data, the electric field is given as $\mathbf{E} = \mathbf{a}_X e^{-0.00067z} e^{-j0.059z}$ V/m. The associated magnetic field is $\mathbf{H} = \mathbf{a}_Y \frac{e^{-0.00067z} e^{-j0.059z}}{133.2 + j1.5}$ A/m. These combine to give the complex Poynting's vector as

$$\mathbf{P} = \frac{\mathbf{E} \times \mathbf{H}^*}{2} = \mathbf{a}_X \frac{e^{-0.00067z} e^{-j0.059z}}{2} \times \mathbf{a}_Y \left(\frac{e^{-0.00067z} e^{-j0.059z}}{133.2 + j1.5} \right)^*$$

$$= \mathbf{a}_Z \frac{e^{-0.00134z}}{2(133.2 - j1.5)} \text{ w/m}^2.$$

Since Poynting's vector is only in the z-direction, there is no contribution to the power surface integral on the constant x and y surfaces. The power out of the cube is given by

$$P_{\text{OUT}} = \oiint_{\text{CUBE}} \mathbf{P} \cdot \mathbf{ds} = \iint_{z=0} \mathbf{P} \cdot \mathbf{ds} + \iint_{z=1} \mathbf{P} \cdot \mathbf{ds}$$

$$= \int_{y=0}^{1} \int_{x=0}^{1} \frac{\mathbf{a}_Z \cdot (-\mathbf{a}_Z) dx dy}{2(133.2 - j1.5)}$$

$$+ \int_{y=0}^{1} \int_{x=0}^{1} \frac{e^{-0.00134} \mathbf{a}_Z \cdot \mathbf{a}_Z dx dy}{2(133.2 - j1.5)}$$

$$= \frac{e^{-0.00134} - 1}{2(133.2 - j1.5)} = -5 - j0.057 \text{ μw}.$$

The average power out of the cube is -5 μw; power flows in the $z = 0$ surface, but less flows out of the $z = 1$ surface due to the losses within the cube. There is a negative reactive power flow out of the volume. There are no sources within the volume so the first term on the RHS is zero. The conduction losses give the power dissipated in the cube as

$$\iiint_{CUBE} \frac{\sigma |\mathbf{E}|^2}{2} dv = \int_{z=0}^{1} \int_{y=0}^{1} \int_{x=0}^{1} \frac{10^{-5} e^{-0.00134z}}{2} dx dy dz$$

$$= 5 \text{ μw}.$$

The power dissipated within the volume is equal to the power lost by the wave. The reactive power within the volume is given by

$$
-j2\omega \iiint\limits_{CUBE} \frac{\mu|\mathbf{H}|^2 - \varepsilon|\mathbf{E}|^2}{4} dv
$$

$$
= -\frac{j4\pi \times 10^6}{4} \iiint\limits_{CUBE} \frac{4\pi \times 10^{-7} e^{-0.00134z}}{|133.2 + j1.5|^2} dxdydz
$$

$$
+ \frac{j4\pi \times 10^6}{4} \iiint\limits_{CUBE} 8(8.854 \times 10^{-12}) e^{-0.00134z} dxdydz
$$

$$
= -j0.057 \ \mu w.
$$

Example 6.8-4. Calculate the DC power flow into a long cylindrical resistor of radius $\rho = a$ via Poynting's vector and with circuit concepts and compare the two results. (Of course, you suspect that the two results should be equal, and we will find that they are.) A uniform current density and electric field exist within the resistor—for a total current I within the resistor. The current density and electric fields are given by $\mathbf{E} = \frac{\mathbf{J}}{\sigma} = \frac{\mathbf{a}_z I}{\pi a^2 \sigma}$ and produce a magnetic field within the resistor given by $\mathbf{H} = \frac{\mathbf{a}_\phi I\rho}{2\pi a^2}$. The power flowing into the resistor can be calculated by the integration of Poynting's vector over the surface of the resistor. Poynting's vector at the cylindrical surface of the resistor is calculated as $\mathbf{P} = \mathbf{E} \times \mathbf{H}^* = \frac{\mathbf{a}_z I}{\pi a^2 \sigma} \times \frac{\mathbf{a}_\phi I^* a}{2\pi a^2} = -\mathbf{a}_\rho \frac{|I|^2}{\sigma 2\pi^2 a^3}$ directed in the negative ρ-direction into the resistor. Note that since these are DC fields, they are real and, hence, $\mathbf{H}^*$ and the peak values of $\mathbf{E}$ and $\mathbf{H}$ are equal to the RMS values and no factor of 2 is used. The total power is calculated as

$$
P_{AVE} = \oiint\limits_{S} \mathbf{P} \cdot \mathbf{ds} = -\int\limits_{z=0}^{L} \int\limits_{\phi=0}^{2\pi} \frac{I^2}{\sigma 2\pi^2 a^3} \mathbf{a}_\rho \cdot \mathbf{a}_\rho a d\phi dz
$$

$$
= -\frac{I^2 L 2\pi}{\sigma 2\pi^2 a^2} = -I^2 \frac{L}{\sigma \pi a^2} = -I^2 R
$$

where the negative sign indicates that the power is into rather than out of the surface of the resistor. The Poynting's vector does not cross the end surfaces so there is no power flow there. An alternative field calculation is the calculation of the power dissipated within the resistor as given by the second term on the RHS of Eq. (6-99) as

$$P_{\text{DISS}} = \iiint_V \sigma |\mathbf{E}|^2 dv = \int_{z=0}^{L} \int_{\phi=0}^{2\pi} \int_{\rho=0}^{a} \sigma \frac{I^2 \rho d\rho d\phi dz}{\pi^2 a^4 \sigma^2}$$

$$= I^2 \frac{\sigma \pi a^2 2\pi L}{2\pi^2 a^4 \sigma^2} = I^2 \frac{L}{\sigma \pi a^2} = I^2 R.$$

Of course the resistance calculations lead to $P_{\text{AVE}} = I^2 R$ as did each of the other two calculations. This is a rather impressive demonstration that the field representations are valid for calculating power dissipated.

6.9 WAVES INCIDENT ON BOUNDARIES

In the discussion so far, the waves have been propagating in unbounded space; there have been no boundaries between different materials. Since the majority of actual waves are incident upon boundaries we must understand wave behavior there. Throughout this section, the media will be assumed to be lossless to simplify the calculations, but the results in lossy media are similar. To begin, consider a linearly polarized, plane wave propagating in the z-direction to be normally incident upon a material boundary located in the $z = 0$ plane as shown in Fig. 6.16.

The electric field of the *incident wave* is expressed as

$$\mathbf{E}^i = \mathbf{a}_X E_o e^{-j\beta_1 z} \tag{6.88}$$

and the corresponding magnetic field as

$$\mathbf{H}^i = \frac{\mathbf{a}_Y E_o e^{-j\beta_1 z}}{\eta_1}. \tag{6.89}$$

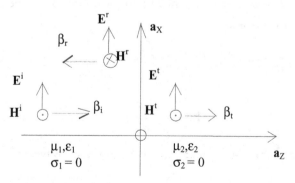

FIGURE 6.16: Plane wave normally incident upon a planar material boundary.

The subscripted β_1 and η_1 denote the values in the region 1, i.e., $z < 0$. Our life experiences with light waves suggest that a wave may be reflected from the material interface and propagates toward the left in region 1. For linear materials, the *reflected wave* will be proportional to the incident wave. More specifically, this relationship is expressed as

$$\Gamma = \left.\frac{E^r}{E^i}\right|_{z=0},\tag{6.90}$$

the ratio of the complex amplitude of the reflected wave to the incident wave at the material boundary. Γ is defined as the *reflection coefficient*; it can be a complex number. Furthermore, since there is only an x-component of the incident wave, there is only an x-component of the reflected wave as well. The reflected electric field is expressed as

$$\mathbf{E}^r = \mathbf{a}_X \Gamma E_o e^{+j\beta_1 z}\tag{6.91}$$

where the exponential term has a $+$ sign since propagation is in the $-z$-direction. The reflected magnetic field is determined from plane wave properties as

$$\mathbf{H}^r = -\mathbf{a}_Y \frac{\Gamma E_o}{\eta_1} e^{+j\beta_1 z}.\tag{6.92}$$

The minus sign can be interpreted as due to $\eta^- = -\eta^+$ or due to the fact that $\mathbf{E}\times\mathbf{H}$ points in the direction of propagation of the wave, $-\mathbf{a}_Z$.

Our experience with light further suggests that a *transmitted wave* may propagate to the right of the boundary. As with the reflected wave, it is polarized as the incident wave and is proportional to it according to

$$T = \left.\frac{E^t}{E^i}\right|_{z=0}.\tag{6.93}$$

T is defined as the *transmission coefficient* similarly to the definition of the reflection coefficient. The transmitted wave is expressed as

$$\mathbf{E}^t = \mathbf{a}_X T E_o e^{-j\beta_2 z}\tag{6.94}$$

and

$$\mathbf{H}^t = \mathbf{a}_Y \frac{T E_o}{\eta_2} e^{-j\beta_2 z}\tag{6.95}$$

where β_2 and η_2 represent the propagation constant and wave impedance in region 2, i.e., $z > 0$.

Waves emanate from a source to the left of the boundary and are reflected from and transmitted through material boundaries as indicated by Eqs. (6.96)–(6.97). This model predicts two waves, the incident and reflected, in region 1, but only one wave, the transmitted, in region 2, since there is no source of a left-propagating wave in this region.

But, we still have two undetermined constants, Γ and T. These can be found by applying the boundary conditions satisfied by all EM fields at the material interface as given by

$$\mathbf{a}_N \times (\mathbf{E}_1 - \mathbf{E}_2) = 0 \qquad (6.96)$$

and

$$\mathbf{a}_N \times (\mathbf{H}_1 - \mathbf{H}_2) = \mathbf{K}. \qquad (6.97)$$

In most cases $\mathbf{K} = 0$ so that the tangential components of both the electric and magnetic fields are equal at the material boundary. In region 1, the electric field is composed of the incident and reflected fields; in region 2, there is only the transmitted field. This leads to

$$\begin{aligned} E_{1\text{TAN}} &= E_o [e^{-j\beta_1 z} + \Gamma e^{+j\beta_1 z}]_{z=0} = E_o(1 + \Gamma) \\ &= E_{2\text{TAN}} = E_o\, T e^{-j\beta_2 z}|_{z=0} = E_o\, T, \end{aligned} \qquad (6.98)$$

which simplifies to

$$1 + \Gamma = T. \qquad (6.99)$$

Similarly, the magnetic field boundary condition gives

$$\begin{aligned} H_{1\text{TAN}} &= E_o \left[\frac{e^{-j\beta_1 z} - \Gamma e^{+j\beta_1 z}}{\eta_1} \right]_{z=0} = E_o \left(\frac{1 - \Gamma}{\eta_1} \right) \\ &= H_{2\text{TAN}} = E_o \frac{T}{\eta_2} e^{-j\beta_2 z}|_{z=0} = E_o \frac{T}{\eta_2} \end{aligned} \qquad (6.100)$$

which simplifies to

$$\frac{1 - \Gamma}{\eta_1} = \frac{T}{\eta_2}. \qquad (6.101)$$

Combining Eqs. (6.100) and (6.101), we obtain solutions for the reflection coefficient as

$$\Gamma = \frac{\eta_2 - \eta_1}{\eta_2 + \eta_1} \qquad (6.102)$$

and the transmission coefficient as

$$T = \frac{2\eta_2}{\eta_2 + \eta_1}. \qquad (6.103)$$

The reflection and transmission of the normally incident plane wave depend only upon the material properties of the two regions. When the materials in region 1 and region 2 are the same, $\Gamma = 0$ and $T = 1$, indicating that there is no reflection at the boundary and all of the incident wave propagates from region 1 into region 2. Of course, this is expected since such a boundary is a mathematical definition, not an actual material boundary.

Most materials are nonmagnetic, i.e., $\mu = \mu_o$, for which the reflection coefficient becomes

$$\Gamma = \frac{\eta_2 - \eta_1}{\eta_2 + \eta_1} = \frac{\sqrt{\dfrac{\mu_o}{\varepsilon_2}} - \sqrt{\dfrac{\mu_o}{\varepsilon_1}}}{\sqrt{\dfrac{\mu_o}{\varepsilon_2}} + \sqrt{\dfrac{\mu_o}{\varepsilon_1}}} = \frac{\sqrt{\varepsilon_1} - \sqrt{\varepsilon_2}}{\sqrt{\varepsilon_1} + \sqrt{\varepsilon_2}}. \tag{6.104}$$

When the dielectric constant of the region on the right is less than the region on the left, the reflection coefficient is positive. The reflected wave will be polarized in the same direction as the incident signal. The transmitted wave will be less than the incident wave (recall that $T = 1 - \Gamma < 1$ for $\Gamma > 0$). In fact, for $\varepsilon_1 \gg \varepsilon_2$ the reflected wave is nearly as large as the incident wave and the transmitted wave is very small. On the other hand, when the dielectric constant of the region on the right is greater than the region on the left, the reflection coefficient is negative. The reflected wave will be oppositely polarized to the incident wave and the transmitted wave will be larger in magnitude than the incident wave.

Since the transmitted wave is larger than the incident wave, does this make sort of an amplifier—a wave amplifier? Intuition suggests that power must be conserved. Let's briefly consider this question. In a lossless medium, we should expect that the power of the incident wave is equal to the sum of the power contained within the reflected and transmitted waves. Alternatively, we should observe the same behavior for the Poynting's vectors for these waves. The Poynting's vector for the incident wave is given by

$$\mathbf{P}^i = \frac{\mathbf{E}^i \mathbf{x} \mathbf{H}^{i*}}{2} = \mathbf{a}_Z \frac{|E_o|^2}{2\eta_1}, \tag{6.105}$$

the Poynting's vector for the reflected wave is

$$\mathbf{P}^r = \frac{\mathbf{E}^r \mathbf{x} \mathbf{H}^{r*}}{2} = -\mathbf{a}_Z \frac{|\Gamma|^2 |E_o|^2}{2\eta_1}, \tag{6.106}$$

and the Poynting's vector for the transmitted wave is

$$\mathbf{P}^t = \frac{\mathbf{E}^t \mathbf{x} \mathbf{H}^{t*}}{2} = \mathbf{a}_Z \frac{|T|^2 |E_o|^2}{2\eta_2}. \tag{6.107}$$

Power conservation predicts that the power into a region that includes the boundary must equal the power out of the region. Choosing a 1-m cube aligned with and centered on the boundary, we can express this in terms of Poynting's vector for the several waves as

$$P_{IN} = |\mathbf{P}^i| = \frac{|E_o|^2}{2\eta_1} = P_{OUT} = |\mathbf{P}^r| + |\mathbf{P}^t|$$

$$= \frac{|E_o|^2}{2} \left(\frac{|\Gamma|^2}{\eta_1} + \frac{|T|^2}{\eta_2} \right) \tag{6.108}$$

or

$$\frac{1}{\eta_1} = \frac{|\Gamma|^2}{\eta_1} + \frac{|T|^2}{\eta_2}$$

$$= \frac{|\eta_2 - \eta_1|^2}{\eta_1 |\eta_2 + \eta_1|^2} + \frac{4\eta_2^2}{\eta_2 |\eta_2 + \eta_1|^2}$$

$$= \frac{1}{\eta_1} \left(\frac{\eta_1^2 + \eta_2^2 - 2\eta_1\eta_2 + 4\eta_1\eta_2}{\eta_1^2 + 2\eta_1\eta_2 + \eta_2^2} \right) = \frac{1}{\eta_1}. \tag{6.109}$$

Indeed, power is conserved at the boundary. The individual electric or magnetic fields associated with the transmitted wave may be greater than their incident wave counterpart. But, their ratio will also be different according to the wave impedance such that the power is conserved.

These arguments of power conservation of waves cannot be applied to waves in lossy media. Though total power is indeed conserved, part of the power of the waves is left within the material in the form of increased material temperature. However, Eq. (6.87) is always valid.

The effects of a single material boundary are completely described by the reflection and transmission coefficients. The propagation constant and the wave impedance depend upon the material properties as well. With these four parameters and frequency, a complete description is given for the fields of plane waves near a boundary.

An interesting feature of the reflection coefficient at a boundary is that it differs in sign depending upon the direction of the incident wave. We have just calculated the reflection for an incident wave from the left, call it Γ_{LEFT}. The calculation for the reflection coefficient for an incident wave from the right requires a reversal of the role of η_1 and η_2 so that $\Gamma_{RIGHT} = -\Gamma_{LEFT}$.

Physicists often write the reflection coefficient for an air–dielectric interface in terms of the *index of refraction* for nonmagnetic dielectric. The index of refraction is defined in terms of the relative permittivity of the dielectric as $n = \sqrt{\varepsilon_R}$.

This analysis has assumed that there is only one boundary present and that the reflected and transmitted waves propagate to $-\infty$ and $+\infty$, respectively. The reflection and transmission coefficients do not completely describe the reflections when a second (or more) boundary is present. A glimpse of this more complicated behavior can be obtained by considering the

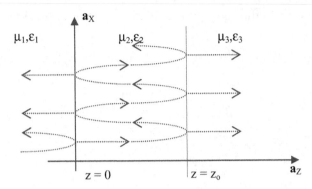

FIGURE 6.17: Multiple reflections from two nearby boundaries.

configuration shown in Fig. 6.17 where a second boundary exists at $z = +z_o$. A qualitative description is that when a wave front is incident upon left boundary at $z = 0$, reflection and transmission takes place as described above. But, when the transmitted wave front impinges upon the right boundary, it acts as an incident wave on this boundary, establishing another reflected and transmitted wave. The reflected wave propagates back toward the left boundary where it acts as an incident wave establishing a third set of reflected and transmitted waves. The reflected wave from this boundary propagates back toward the right boundary and the process continues ad infinitum. An infinite number of waves are moving to the right and to the left in the region $0 < z < z_o$; an infinite number of waves are moving to the left in the region $z < 0$; an infinite number of waves are moving to the right in the region $z > z_o$. The sum of these waves establishes steady-state reflected waves from the two boundaries to the left and steady-state transmitted waves from the two boundaries to the right. But, the calculation of the "reflection" and "transmission" coefficients for this situation is much more complicated. We will delay a quantitative analysis of this case till our discussion of transmission lines.

Finally, let's take a brief look at obliquely incident waves on a boundary, see Fig. 6.18. A wave incident from the left travels in a direction defined by $\mathbf{a}_\beta = \mathbf{a}_X \sin \theta_I + \mathbf{a}_Z \cos \theta_i$. Since

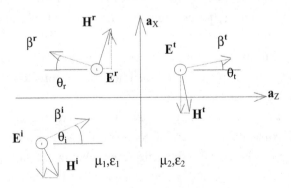

FIGURE 6.18: Obliquely incident wave on a material boundary.

the electric fields for TEM waves propagating in this direction must be perpendicular to this vector, they can be decomposed into components either parallel or perpendicular to the *plane of incidence*. The plane of incidence is defined as the plane formed by $\mathbf{a}_\beta$ and the normal to the material boundary, in this case the $y = 0$ plane. Let's consider the perpendicular polarization which results in the electric field of the incident wave expressed as

$$\mathbf{E}^i = \mathbf{a}_Y E_\perp e^{-j\beta_1(x\sin\theta_i + z\cos\theta_i)}. \qquad (6.110)$$

The associated magnetic field must be perpendicular to the electric field and the direction of propagation, so it lies in the $x-z$ plane as shown in Fig. 6.18. The components of the magnetic field are determined from the geometry, the magnitude from the electric field and the wave impedance, so that

$$\mathbf{H}^i = (-\mathbf{a}_X \cos\theta_i + \mathbf{a}_Z \sin\theta_i)\frac{E_\perp}{\eta_1} e^{-j\beta_1(x\sin\theta_i + z\cos\theta_i)}. \qquad (6.111)$$

Similar to the normal incidence, the electric field of the reflected wave is expressed as

$$\mathbf{E}^r = \mathbf{a}_Y \Gamma_\perp E_\perp e^{-j\beta_1(x\sin\theta_r - z\cos\theta_r)} \qquad (6.112)$$

and the corresponding magnetic field as

$$\mathbf{H}^r = (\mathbf{a}_X \cos\theta_r + \mathbf{a}_Z \sin\theta_r)\frac{\Gamma_\perp E_\perp}{\eta_1} e^{-j\beta_1(x\sin\theta_r - z\cos\theta_r)}. \qquad (6.113)$$

Note the reversed direction of the z-directed propagation in the exponent and the oppositely directed x-component of the field to satisfy the TEM property of the wave. Finally, the transmitted wave is formed as

$$\mathbf{E}^t = \mathbf{a}_Y T_\perp E_\perp e^{-j\beta_2(x\sin\theta_t + z\cos\theta_t)} \qquad (6.114)$$

and

$$\mathbf{H}^t = (-\mathbf{a}_X \cos\theta_t + \mathbf{a}_Z \sin\theta_t)\frac{T_\perp E_\perp}{\eta_2} e^{-j\beta_2(x\sin\theta_t + z\cos\theta_t)} \qquad (6.115)$$

where the incident, reflected, and transmitted angles have been assumed to be independent. Invoking the electric field boundary condition at $z = 0$, we obtain

$$E_\perp e^{-j\beta_1 x\sin\theta_i} + \Gamma_\perp E_\perp e^{-j\beta_1 x\sin\theta_r} = T_\perp E_\perp e^{-j\beta_2 x\sin\theta_t}. \qquad (6.116)$$

Since this must be true for all values of x along the boundary, the phase terms which depend upon x must all be equal, i.e.,

$$\beta_1 \sin\theta_i = \beta_1 \sin\theta_r = \beta_2 \sin\theta_t. \qquad (6.117)$$

The first equality is satisfied when

$$\theta_r = \theta_i. \tag{6.118}$$

This is *Snell's law* equating incident and reflected angles. The equality of the first and third terms is satisfied when

$$\sin \theta_t = \frac{\beta_1 \sin \theta_i}{\beta_2} = \frac{\sqrt{\mu_1 \varepsilon_1} \sin \theta_i}{\sqrt{\mu_2 \varepsilon_2}}, \tag{6.119}$$

Snell's law relating incident and transmitted angles. Most materials are nonmagnetic for which $\mu_1 = \mu_2 = \mu_o$ and Snell's law becomes

$$\sin \theta_t = \frac{\sqrt{\varepsilon_1} \sin \theta_i}{\sqrt{\varepsilon_2}}. \tag{6.120}$$

When $\varepsilon_1 > \varepsilon_2$, $\theta_t = \pi/2$ for values of $\theta_I < \pi/2$. This describes what is known as the *critical angle* and describes the largest angle of incidence for which a transmitted wave propagates into region 2, i.e.,

$$\theta_{\text{CRIT}} = \sin^{-1}\left(\sqrt{\frac{\varepsilon_2}{\varepsilon_1}}\right). \tag{6.121}$$

This angle of incidence produces a transmitted wave parallel to the boundary; greater angles of incidence produce complex values of transmitted angle and much more complicated results (reserved for an advanced course). This effect occurs when looking out of a swimming pool. The water is region 1 and the air is region 2 with $\varepsilon_2 = \varepsilon_o < \varepsilon_1 = 81\varepsilon_o$.

With the exponential terms of Eq. (6.116) equal, the electric field boundary condition becomes

$$1 + \Gamma_\perp = T_\perp. \tag{6.122}$$

The magnetic field boundary condition (in the absence of surface current density, **K**) becomes

$$-\cos\theta_i \frac{E_\perp}{\eta_1} + \cos\theta_r \frac{\Gamma_\perp E_\perp}{\eta_1} = -\cos\theta_t \frac{T_\perp E_\perp}{\eta_2} \tag{6.123}$$

or rewritten as

$$-\frac{\cos\theta_i}{\eta_1} + \frac{\Gamma_\perp \cos\theta_r}{\eta_1} = -\frac{T_\perp \cos\theta_t}{\eta_2}. \tag{6.124}$$

Combining Eqs. (6.122) and (6.124), we obtain the solutions for $\Gamma_\perp$ and $T_\perp$ as

$$\Gamma_\perp = \frac{\eta_2 \cos\theta_i - \eta_1 \cos\theta_t}{\eta_2 \cos\theta_i + \eta_1 \cos\theta_t} \tag{6.125}$$

and

$$T_\perp = \frac{2\eta_2 \cos\theta_i}{\eta_2 \cos\theta_i + \eta_1 \cos\theta_t} \qquad (6.126)$$

where θ_i and θ_t are related by Eq. (6.119). These forms are more complicated due to the oblique incidence. However, they are identical to their counterparts for normal incidence. If the angle of incidence becomes 0, then they reduce to the results obtained for normal incidence as we expect. Though time prevents us from looking at this form in more detail, the essential characteristics of the wave behavior are similar to the normal case. The results for the electric field parallel to the plane of incidence can be derived in the same manner, but that must wait until a homework problem.

Example 6.9-1. Calculate the percentage of the incident power that is reflected and that is transmitted when a plane wave is normally incident upon the surface of a calm lake. The relative permittivity of water is 81 so the reflection coefficient is given by Eq. (6.102) as

$$\Gamma = \frac{\eta_2 - \eta_1}{\eta_2 + \eta_1} = \frac{\sqrt{\dfrac{\mu_o}{81\varepsilon_o}} - \sqrt{\dfrac{\mu_o}{\varepsilon_o}}}{\sqrt{\dfrac{\mu_o}{81\varepsilon_o}} + \sqrt{\dfrac{\mu_o}{\varepsilon_o}}} = \frac{\sqrt{\varepsilon_o} - \sqrt{81\varepsilon_o}}{\sqrt{\varepsilon_o} + \sqrt{81\varepsilon_o}}$$

$$= \frac{1-9}{1+9} = -0.8$$

and the transmission coefficient is given by Eq. (6.103) as

$$T = \frac{2\eta_2}{\eta_2 + \eta_1} = \frac{2\sqrt{\dfrac{\mu_o}{81\varepsilon_o}}}{\sqrt{\dfrac{\mu_o}{81\varepsilon_o}} + \sqrt{\dfrac{\mu_o}{\varepsilon_o}}} = \frac{\dfrac{2}{9}}{\dfrac{1}{9} + 1} = 0.2.$$

The incident power density is $\mathbf{P}^i = |\mathbf{E}_o|^2/2\eta_1$; the reflected power density is $\mathbf{P}^r = |\Gamma|^2|\mathbf{E}_o|^2/2\eta_1$; the transmitted power density is $\mathbf{P}^t = |T|^2|\mathbf{E}_o|^2/2\eta_2$. The fraction of incident power that is reflected is $\mathbf{P}^r/\mathbf{P}^i = |\Gamma|^2 = 0.64$; the fraction of the incident power that is transmitted is $\mathbf{P}^t/\mathbf{P}^i = |T|^2\eta_1/\eta_2 = (0.04)(1/9) = 0.36$. The two fractions combine to equal the total incident power!

Example 6.9-2. Compare the reflection coefficients of a normally incident plane wave in air of frequency 100 MHz upon a PEC and upon copper. The wave impedance of a good conductor is given by $\eta = [j\omega\mu/\sigma]^{1/2}$ so $\eta_{CU} = 0.00369e^{j\pi/4}$ and $\eta_{PEC} = 0$. This leads to $\Gamma_{CU} = (0.00369e^{j\pi/4} - 377)/(0.00369e^{j\pi/4} + 377) = -0.999986 + j0.0000138$ and $\Gamma_{PEC} = -1$. The difference between copper and PEC is insignificant for most reflection coefficient calculations.

Example 6.9-3. Your pet piranha looks hungrily out of the tank at you. What range of angles must she scan to see the entire room? Let's assume the glass has negligible effect. The permittivity in the water (region 1) is $81\varepsilon_o > \varepsilon_o$. Therefore, the critical angle inside the tank at which she can see things at a $90°$ angle outside of her tank wall is given by Eq. (6.121) as $\theta_{CRIT} = \sin^{-1}\left(\sqrt{\frac{\varepsilon_2}{\varepsilon_1}}\right) = \sin^{-1}\left(\sqrt{\frac{\varepsilon_o}{81\varepsilon_o}}\right) = \sin^{-1}\left(\frac{1}{9}\right) = 6.4°$. So over a range of $12.8°$, she can scan the entire $180°$ outside of one wall of her tank.

6.10   STANDING WAVES

When a wave is incident normally upon a boundary as in Fig. 6.19, there are two fields in region 1, the incident and the reflected. The total field in region 1 is expressed as

$$\mathbf{E}_1 = \mathbf{a}_X E_o \left(e^{-j\beta z} + \Gamma e^{+j\beta z}\right), \qquad (6.127)$$

a traveling wave to the right and one to the left. Some purely mathematical manipulations put this into a form that provides an alternate interpretation,

$$\begin{aligned} \mathbf{E}_1 &= \mathbf{a}_X E_o \left(e^{-j\beta z} + \Gamma e^{+j\beta z}\right) \\ &= \mathbf{a}_X E_o \left([1 - \Gamma]e^{-j\beta z} + \Gamma[e^{-j\beta z} + e^{+j\beta z}]\right) \\ &= \mathbf{a}_X E_o \left(\underbrace{[1 - \Gamma]e^{-j\beta z}}_{\text{TRAVELING WAVE}} + \underbrace{2\Gamma\cos\beta z}_{\text{STANDING WAVE}}\right). \end{aligned} \qquad (6.128)$$

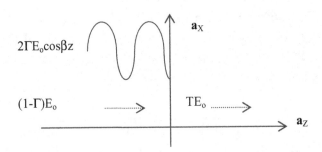

FIGURE 6.19: Standing wave.

The first term represents a portion of the incident wave moving in the $+z$-direction. That portion of the incident wave that is reflected is regrouped "mathematically" with the reflected wave to form a standing wave represented by the last term. It is actually composed of two, equal, but oppositely directed waves that combine to form a single *standing wave*. Since the combined result has only amplitude (but no phase) variation with z, it is not a propagating wave. When transformed to the time domain, it is just a cosine spatial distribution that varies in magnitude. This term is similar to the standing waves seen on vibrating strings, with nonmoving positions at which the wave is peaked and others at which it is nulled. However, for the vibrating string, the reflection coefficient is effectively unity so there is no propagating wave.

The actual observation of a standing electromagnetic wave is not as easy as with the vibrating string, but it is possible to measure properties of the total field as represented by Eq. (6.127). The maxima and minima of the field can be readily measured and from these values information about the reflection coefficient can be determined. Consider the two terms of the total field as phasors; the maximum value of their sum occurs when they have the same phase angle so that their magnitudes can be added to give

$$|\mathbf{E}_1|_{\text{MAX}} = |E_o|(1 + |\Gamma|). \tag{6.129}$$

Similarly, the minimum value of the field is given by

$$|\mathbf{E}_1|_{\text{MIN}} = |E_o|(1 - |\Gamma|) \tag{6.130}$$

where the fact that $|\Gamma| \leq 1$ has been used. The value $|E_o|$ can be eliminated by taking the quotient to obtain the *standing wave ratio*, SWR, as

$$SWR = \frac{|\mathbf{E}_1|_{\text{MAX}}}{|\mathbf{E}_1|_{\text{MIN}}} = \frac{1 + |\Gamma|}{1 - |\Gamma|}. \tag{6.131}$$

The ratio of the maximum to minimum field values depends upon the magnitude of the reflection coefficient. This relatively simple measurement enables an indication of the magnitude of the reflected signal as

$$|\Gamma| = \frac{SWR - 1}{SWR + 1}. \tag{6.132}$$

For reflection coefficients that approach unity in magnitude, the SWR becomes very large, approaching infinity. The actual signal doesn't approach infinity, rather, the minimum approaches zero so that the quotient becomes large. A small SWR is usually desired as it means less signal is reflected. This is a very common measurement for transmission lines and we will use this a great deal in the next chapter.

Example 6.10-1. Calculate the SWR for the incident and reflected waves of Example 6.9-1. The SWR is calculated easily as $SWR = \frac{1+|\Gamma|}{1-|\Gamma|} = \frac{1+0.8}{1-0.8} = 9$. This means that the maximum electric field intensity is 9 times the minimum electric field intensity in the region above the lake.

Though there is much more to learn about electromagnetic waves, we have covered the fundamentals and must move to a more specialized version of electromagnetic waves—those that are guided by metallic wires. These are known as transmission lines; they come in many forms. Voltages, currents, and impedances can be measured. Accordingly, we will be able to use a mixture of circuit and wave concepts to describe transmission line behavior in the next chapter.

CHAPTER 7

Transmission Lines

7.1   GUIDED TEM WAVES

The propagation of TEM electromagnetic waves in space is the basis for many communication systems in common use today. Worldwide and extraterrestrial communications between widely dispersed and/or mobile stations such as cellular phones, satellite communications, and collision avoidance radar utilize this phenomenon. However, since these waves cannot be limited to well-defined spatial regions, their signals are not secure from detection by other users. Moreover, they may interfere with other signals. There are many applications such as telephony, cable TV, and LANs where it is desired to guide the signals from one site to other selected, stationary sites. Transmission lines offer low-cost, convenient, secure, and easily controlled propagation media for these applications. From another perspective, circuit board traces between lumped elements can behave often as an unintended transmission line. Waves on one trace can induce an unwanted signal onto another trace. We need to understand transmission lines!

A very popular form of electromagnetic transmission system is fiber optics. This media is most effective at optical frequencies where the guide can be made very small, less than millimeters in diameter. With carrier frequencies in the optical frequency range, a lot of wideband data transmissions are possible. The dielectric properties of the guide keep the waves confined mainly within the fiber with very little leakage. However, we will concentrate upon lower frequency transmission media that utilize two metallic conductors separated by a dielectric insulator.

The application of a voltage to the terminals of a two-wire transmission line establishes electromagnetic fields in the vicinity of the metallic conductors. These electromagnetic fields propagate along the transmission line with a finite velocity, establishing a voltage–current pair that propagates along the transmission line as well. The familiar impedance concepts are useful in analyzing these voltage and current waves.

Though transmission lines take on many different forms, they all share several common features. Firstly, they use two insulated metallic conductors. The shape or configuration is relatively unimportant to proper operation and is often chosen for convenience of the application. Secondly, the cross-sectional dimensions are small compared to the wavelength of the highest frequency transmitted on the line. When the cross section is of the order of a wavelength, other modes of propagation become possible; operation in this manner is to be avoided. Thirdly, they are relatively long in the axial direction along which the waves propagate. In fact, the axial dimension or length of the line is usually considered to be infinite for analysis purposes. Finally, there is no variation of the cross-sectional dimensions with length. Dimensional changes in cross sections alter the propagation characteristics of the line and complicate the solutions. Whenever transmission lines are bent around the corners or nearby objects, variations in the cross section are introduced, but these effects are minor as long as the bends are gradual. For simplicity, we will often assume that the metallic conductors can be modeled as PECs. For more accurate calculations, actual conductor properties must be used. Multi-stranded and silver-plated conductors are often used to minimize losses. In addition, we will assume that the dielectrics are lossless which is reasonable for all but the most exacting of applications.

The most inexpensive form of transmission line is known as *twin-lead*; see Fig. 7.1(a). Two metallic conductors, usually the same size, are embedded in a low-loss dielectric material. Since the two wires are not shielded, unwanted external magnetic fields can penetrate the region between the two wires and induce interfering voltages. Due to reciprocity, the fields of twin-lead transmission lines can induce voltages in nearby wires or other twin-lead lines as well. In addition, the presence of nearby materials can alter the fields surrounding the conductors and change the propagation characteristics of the line. To partially overcome this weakness, *twisted-pair* transmission lines are often used, particularly in low-speed LAN applications; see Fig. 7.1(b).

Isolation of the transmission line fields and external fields is readily accomplished with *coaxial* transmission lines (often called *coax*); see Fig. 7.1(c). The outer conductor completely

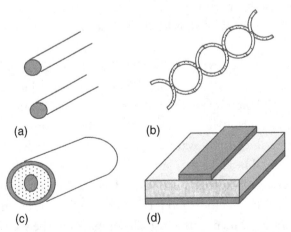

FIGURE 7.1: Transmission lines: (a) twin-lead, (b) twisted-pair, (c) coaxial, and (d) microstrip.

surrounds the inner conductor and the region containing the transmission line fields. Only a slight amount of leakage of the internal fields through the outer conductor occurs. Likewise, the outer conductor shields the internal fields from the effects of external fields and materials. These excellent properties of coaxial cable cost more than twin-lead transmission lines. In addition, coaxial lines have more loss unless materials are carefully selected.

Microstrip transmission lines are especially useful in high frequency circuits; see Fig. 7.1(d). A double-sided, copper-printed circuit board is used for such lines. One copper side remains intact to approximate an infinite ground plane; the other conductor is etched to the desired width between the desired points. The two insulated conductors form a transmission line. Adjacent conductor traces on a digital backplane often can behave in a similar way to inadvertently form a transmission line. Microstrip lines are not shielded, but have very little gap between the two conductors, so are relatively immune to external fields.

The remainder of this chapter develops the generalized theory and application of all types of transmission lines.

7.2 TEM WAVES ON TRANSMISSION LINES

The DC electric fields within a coaxial cable are particularly simple, i.e., they are angularly symmetric, show radial dependence only, and are radially directed as in Example 1.29-1. It seems reasonable that very low frequency fields should exhibit spatial variations in a transverse plane very similar to DC fields. The magnetic field is angularly directed in the transverse plane as well. Furthermore, we attribute the delay of a signal from one end of a coax to the other to wave propagation along the axis of the coax. This suggests that the axial dependence of time-varying fields within a coax must show a wave dependence such as in plane waves, i.e., $e^{-j\beta z}$, for a wave in the $+z$-direction. If a product form of the electromagnetic field is assumed, then the phasor form of the electric field within a coax is represented as the product of the DC field's transverse dependence and the axial propagation characteristics as

$$\mathbf{E} = \frac{V_O e^{-j\beta z}}{\ln\left(\frac{b}{a}\right)\rho}\mathbf{a}_\rho \qquad (7.1)$$

where V_o is the voltage difference between the inner and outer conductor. Similar reasoning leads to the magnetic field as

$$\mathbf{H} = \frac{I_o e^{-j\beta z}}{2\pi\rho}\mathbf{a}_\phi \qquad (7.2)$$

where I is the current flowing in the center conductor and returning via the outer conductor, see Fig. 7.2.

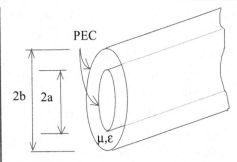

FIGURE 7.2: Coaxial transmission line.

Since these fields are TEM to the assumed axial direction of propagation in a manner similar to plane waves in space, it seems plausible that they describe TEM waves within a coax line. If the coax contains only homogeneous material and no sources between the electrodes, then these fields must satisfy the Helmholtz wave equation

$$\nabla^2 \mathbf{E} + \omega^2 \mu \varepsilon \mathbf{E} = 0. \qquad (7.3)$$

Substituting Eq. (7.1) into the cylindrical form of Eq. (7.3), we obtain

$$
\begin{aligned}
&\nabla^2 \mathbf{E} + \omega^2 \mu \varepsilon \mathbf{E} \\
&= \mathbf{a}_\rho \left(\frac{\partial}{\rho \partial \rho} \left(\rho \frac{\partial E_\rho}{\partial \rho} \right) + \frac{\partial^2 E_\rho}{\rho^2 \partial \phi^2} + \frac{\partial^2 E_\rho}{\partial z^2} + \omega^2 \mu \varepsilon E_\rho \right) \\
&= \mathbf{a}_\rho \left(\frac{\partial^2 E_\rho}{\partial z^2} + \omega^2 \mu \varepsilon E_\rho \right) = \mathbf{a}_\rho \left(-\beta^2 + \omega^2 \mu \varepsilon \right) E_\rho = 0 \qquad (7.4)
\end{aligned}
$$

which has solutions of

$$\beta = \pm \omega \sqrt{\mu \varepsilon}. \qquad (7.5)$$

The assumed TEM plane wave solution for the electric field is valid within the coax as well as in free space! The electric field is given by

$$\mathbf{E} = \frac{V_O e^{-j\omega\sqrt{\mu\varepsilon}z}}{\ln\left(\frac{b}{a}\right)\rho} \mathbf{a}_\rho \qquad (7.6)$$

for a wave propagating in the $+z$-direction. Using Faraday's law, we are able to calculate the accompanying magnetic field as

$$\mathbf{H} = \frac{\nabla \times \mathbf{E}}{-j\omega\mu} = \frac{-j\beta}{-j\omega\mu} \frac{V_O e^{-j\omega\sqrt{\mu\varepsilon}z}}{\ln\left(\frac{b}{a}\right)\rho} \mathbf{a}_\phi = \frac{E_\rho}{\sqrt{\frac{\mu}{\varepsilon}}} \mathbf{a}_\phi \qquad (7.7)$$

The electric and magnetic waves within a coax are related by the wave impedance similar to waves in space.

A closer look at Eq. (7.4) reveals that the first two terms (with ρ and ϕ partial derivatives) vanish since they are of the form of DC solutions and, thus, satisfy the two-dimensional Laplace's equation. They represent the spatial variation of the voltage throughout a constant z-plane, i.e., a transverse cross section. This predicts the same transverse spatial behavior of

DC and AC fields in all types of transmission lines for frequencies with wavelengths longer than about twice the diameter of the coax. Details of this phenomenon are deferred to more advanced study of electromagnetics.

This type of field behavior is valid for other transmission line configurations as well, though the mathematics is often not as simple. In general, the electromagnetic field on a transmission line is given by

$$\mathbf{E} = V_O(x_{t1}, x_{t2})e^{-j\omega\sqrt{\mu\varepsilon}z}\mathbf{a}_{t1} \qquad (7.8)$$

and

$$\mathbf{H} = \frac{V_O(x_{t1}, x_{t2})e^{-j\omega\sqrt{\mu\varepsilon}z}}{\eta}\mathbf{a}_{t2} \qquad (7.9)$$

where $V_o(x_{t1}, x_{t2})$ is the solution of Laplace's equation in the transverse plane for the cross-sectional geometry of the transmission line, x_{t1} and x_{t2} are coordinates in the transverse plane, and $\mathbf{a}_{t1}$ and $\mathbf{a}_{t2}$ are perpendicular unit vectors in the transverse plane.

Power flow down the transmission line can be calculated by the use of Poynting's vector as

$$P_{AVE} = Re\left(\iint\limits_{\substack{\text{Cross} \\ \text{Section}}} \mathbf{P}\cdot\mathbf{ds}\right) = Re\left(\iint\limits_{\substack{\text{Cross} \\ \text{Section}}} \frac{\mathbf{E}\times\mathbf{H}^*}{2}\cdot\mathbf{ds}\right)$$

$$= Re\left(\iint\limits_{\substack{\text{Cross} \\ \text{Section}}} \frac{|V_o(x_{t1}, x_{t2})|^2}{2\eta}dx_{t1}dx_{t2}\right). \qquad (7.10)$$

7.3   VOLTAGE AND CURRENT WAVES

It is comforting to know that waves on a transmission line are similar to those in free space, but it would be more convenient if we could represent them in terms of measurable voltages and currents. Through the use of Maxwell's equations in integral form, the transmission line behavior can be represented in terms of the voltage and current. Consider the section of transmission line shown in Fig. 7.3. The transmission line is composed of two, parallel PEC conductors of different shape and size. Sufficient accuracy is obtained if we assume that the currents are uniformly distributed throughout the conductors with current densities $\mathbf{J}_1$ and $\mathbf{J}_2$, respectively. The total current in the two wires is the same, $I(z)$, but these are oppositely directed. A voltage difference between the two conductors, $V(z)$, establishes an electric field $\mathbf{E}$. The voltage and current are functions of axial position due to the waves on the line.

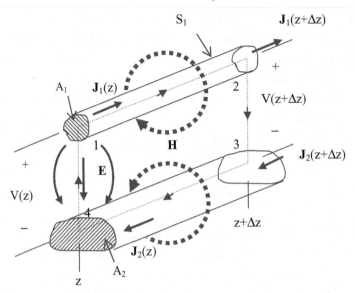

FIGURE 7.3: General transmission line.

Note that the electric field is directed from the more positive upper conductor toward the more negative lower conductor. The magnetic field encircles the two conductors in opposite directions due to the oppositely directed currents, but is totally directed into the plane between the two conductors.

Application of Faraday's law to the rectangular loop defined in the direction of points 1-2-3-4-1 gives

$$\oint_{4-1-2-3-4} \mathbf{E} \cdot \mathbf{dl} = -j\omega \iint_{1234} \mathbf{B} \cdot \mathbf{ds}. \tag{7.11}$$

The LHS is evaluated as

$$\oint_{4-1-2-3-4} \mathbf{E} \cdot \mathbf{dl} = \int_{4}^{1} \mathbf{E} \cdot \mathbf{dl} + \int_{1}^{2} \mathbf{E} \cdot \mathbf{dl} + \int_{2}^{3} \mathbf{E} \cdot \mathbf{dl} + \int_{3}^{4} \mathbf{E} \cdot \mathbf{dl}$$

$$= -V(z) + 0 + V(z + \Delta z) + 0$$

$$\approx -V(z) + V(z) + \frac{\partial V(z)}{\partial z} \Delta z + \cdots$$

$$= \frac{\partial V(z)}{\partial z} \Delta z \tag{7.12}$$

where the integrals from 1-2 and 3-4 vanish because they are within the PEC conductors where the electric field is zero and a Taylor's series expansion of $V(z + \Delta z)$ has been used. The RHS of Faraday's law is expressed as

$$-j\omega \iint\limits_{1234} \mathbf{B} \cdot \mathbf{ds} = -j\omega I(z) \left(\frac{\iint\limits_{1234} \mu \mathbf{H} \cdot \mathbf{ds}}{I(z)} \right)$$

$$= -j\omega I(z) \left\{ \frac{\Delta L}{\Delta z} \right\} \Delta z = -j\omega I(z) \mathcal{L} \Delta z \qquad (7.13)$$

where the incremental inductance of the Δz section is defined as $\Delta L = \Delta \Psi_m / I$ and the *inductance/meter* is defined as $\mathcal{L} = \Delta L / \Delta z$. Equations (7.12) and (7.13) combine to give

$$\frac{\partial V(z)}{\partial z} = -j\omega \mathcal{L} I(z). \qquad (7.14)$$

The axial rate of change of voltage drop is proportional to the current on the line. Additional properties of the transmission line can be gained by application of charge conservation to the upper conductor as

$$\oiint\limits_{S_1} \mathbf{J} \cdot \mathbf{ds} = -j\omega Q. \qquad (7.15)$$

The LHS represents the net current flow into the region Δz section of the upper conductor expressed as

$$\oiint\limits_{S_1} \mathbf{J} \cdot \mathbf{ds} = -\iint\limits_{A_1} \mathbf{J}(z) \cdot \mathbf{ds} + \iint\limits_{A_1} \mathbf{J}(z + \Delta z) \cdot \mathbf{ds}$$

$$= -I(z) + I(z + \Delta z)$$

$$\approx -I(z) + I(z) + \frac{\partial I(z)}{\partial z} \Delta z + \cdots = \frac{\partial I(z)}{\partial z} \Delta z \qquad (7.16)$$

where a Taylor's series expansion has been used to simplify results. The RHS represents the charge that resides on the incremental section of upper conductor as

$$-j\omega Q = -j\omega \oiint\limits_{S_1} \varepsilon \mathbf{E} \cdot \mathbf{ds} = -j\omega \left(\frac{\oiint\limits_{S_1} \mathbf{D} \cdot \mathbf{ds}}{V(z)} \right) V(z)$$

$$= -j\omega V(z) \left\{ \frac{\Delta C}{\Delta z} \right\} \Delta z = -j\omega \mathcal{C} V(z) \Delta z \qquad (7.17)$$

where the incremental capacitance of the Δz section is defined as $\Delta C = \Delta \Psi_e / V$ and the *capacitance/meter* is defined as $\mathcal{C} = \Delta C / \Delta z$. Equations (7.16) and (7.17) combine to give

$$\frac{\partial I(z)}{\partial z} = -j\omega\mathcal{C}V(z); \tag{7.18}$$

the axial rate of change of current on the line is proportional to the voltage drop.

With Eqs. (7.13) and (7.18), transmission line behavior, originally expressed in terms of electromagnetic fields, is now expressed in terms of voltages and currents. These equations are known as *telegrapher's equations*, first derived by Oliver Heaviside to predict the performance of the first trans-Atlantic cable. The next section discusses these equations in detail.

Example 7.3-1. Calculate the inductance/meter and the capacitance/meter for the coaxial transmission line of Fig. 7.2. Either DC or AC transverse fields can be used for these calculations; let's use the DC form. The magnetic field of a coaxial transmission line is calculated by using Ampere's law as $\mathbf{B} = \mu\mathbf{H} = \mathbf{a}_\phi \frac{\mu I}{2\pi\rho}$. The total flux enclosed by the currents for 1 meter length of line is calculated as $\Psi_m = \iint_A \mathbf{B} \cdot \mathbf{ds} = \int_{z=0}^{1} \int_{\rho=a}^{b} \frac{\mu I}{2\pi\rho}\mathbf{a}_\phi \cdot \mathbf{a}_\phi d\rho dz = \frac{\mu I}{2\pi} \ln\left(\frac{b}{a}\right)$. The inductance/meter is given by $\mathscr{L} = \frac{\Psi_m}{I} = \frac{\mu}{2\pi} \ln\left(\frac{b}{a}\right)$ H/m. The electric field between the two conductors is $\mathbf{D} = \varepsilon\mathbf{E} = \mathbf{a}_\rho \frac{\varepsilon V_o}{\ln(b/a)\rho}$ so that the total electric flux from the center conductor to the outer conductor for 1 meter length of line is $\Psi_e = \iint_A \mathbf{D} \cdot \mathbf{ds} = \int_{z=0}^{1} \int_{\phi=0}^{2\pi} \frac{\varepsilon V_o}{\ln(b/a)\rho}\mathbf{a}_\rho \cdot \mathbf{a}_\rho \rho d\phi dz = \frac{2\pi\varepsilon V_o}{\ln(b/a)}$. The capacitance/meter is given by $\mathcal{C} = \frac{\Psi_e}{V} = \frac{2\pi\varepsilon}{\ln(b/a)}$ F/m.

7.4   TELEGRAPHER'S EQUATIONS AND DISTRIBUTED ELEMENT CIRCUIT MODELS

Telegrapher's equations

$$\frac{\partial V(z)}{\partial z} = -j\omega\mathscr{L}I(z)$$

$$\frac{\partial I(z)}{\partial z} = -j\omega\mathcal{C}V(z) \tag{7.19}$$

provide a simple, scalar description of TEM wave phenomena on transmission lines. The effects of the transmission line geometry and material are contained within the parameters of inductance/meter and capacitance/meter. These are referred to as distributed elements since they are not localized as a lumped element but extend the entire axial length of the line. But, they can be used to provide a circuit model if we consider an incremental section of line Δz as shown in Fig. 7.4. The total inductance and capacitance present in this section are $\Delta L = \mathscr{L}\Delta z$

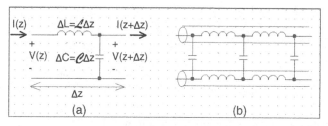

FIGURE 7.4: Incremental transmission line model: (a) electric model and (b) geometric–electric relationship.

and $\Delta C = \mathcal{C}\Delta z$, respectively. Multiplication of telegrapher's equations by Δz leads to

$$\frac{\partial V(z)}{\partial z}\Delta z = -j\omega\mathcal{L}\Delta z I(z) \Rightarrow \Delta V(z) = -j\omega\Delta L I(z)$$

$$\frac{\partial I(z)}{\partial z}\Delta z = -j\omega\mathcal{C}\Delta z V(z) \Rightarrow \Delta I(z) = -j\omega\Delta C V(z). \qquad (7.20)$$

The change in the voltage drop across the transmission line is due to an inductive voltage drop caused by the transmission line current. This is modeled by a series incremental inductance through which the transmission line current flows. Similarly, the change in current in the transmission line is caused by a displacement current flow due to the transmission line voltage drop. This is modeled by a shunt incremental capacitance across which the transmission line voltage drop is impressed. These circuit approximations are valid as long as the incremental section is small compared to wavelength, i.e., $\Delta z \ll \lambda$. If this condition is not met, then the electromagnetic field equations must be used to describe the situation. The combination of these two effects leads to the incremental circuit model for a transmission line as shown in Fig. 7.4(a).

The model has a physical basis as well. The flow of current has an associated magnetic energy storage that accounts for the inductive term; the voltage drop between two insulated conductors has an associated electric energy storage that accounts for the capacitive term. The physical model is shown in Fig. 7.4(b). The inductance and capacitance are distributed everywhere along the transmission lines. But, it is more convenient to represent them in the series-shunt circuit model of Fig. 7.4(a). Several other forms of circuit models, shown in Fig. 7.5, all lead to the same telegrapher's equations. Unless otherwise noted, we will use the circuit model of Fig. 7.4(a) throughout this text.

Application of KVL to the model of Fig. 7.4(a) leads to

$$V(z) + j\omega\Delta L I(z) = V(z + \Delta z) \qquad (7.21)$$

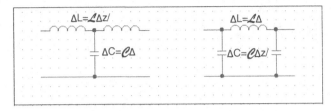

FIGURE 7.5: Alternate transmission line lumped circuit models.

as we obtained from Faraday's law. Similarly, KCL applied to the RH node of Fig. 7.4(a) leads to

$$I(z) = j\omega \Delta C V(z) + I(z + \Delta z), \tag{7.22}$$

in agreement with the conservation of charge.

The simplicity offered by the scalar circuit models makes them much more attractive than the vector fields equations. In addition, they provide the basis for using the circuit concept of impedance, as well.

Confirmation that this model describes the waves on a transmission line is obtained by taking the axial derivative of the circuit form of Faraday's law,

$$\frac{\partial^2 V(z)}{\partial z^2} = -j\omega\mathcal{L}\frac{\partial I(z)}{\partial z} = -j\omega\mathcal{L}[-j\omega\mathcal{C}V(z)]$$
$$= -\omega^2\mathcal{L}\mathcal{C}V(z). \tag{7.23}$$

This is the scalar form of the Helmholtz wave equation in terms of circuit parameters where

$$\beta = \pm\omega\sqrt{\mathcal{L}\mathcal{C}}, \tag{7.24}$$

the positive sign is used for the propagation in the negative z-direction and the negative sign is used for the propagation in the positive z-direction. The propagation constant depends directly on the inductance and capacitance/meter of the line. Indirectly, this reduces to the same dependence upon μ and ε shown by plane waves since both distributed parameters depend upon material properties and are defined for the same transmission line configuration. Of course, velocity of propagation and wavelength are similarly defined as

$$v_P = \frac{\omega}{\beta} = \pm\frac{1}{\sqrt{\mathcal{L}\mathcal{C}}} \tag{7.25}$$

and

$$\lambda = \frac{2\pi}{\beta} = \frac{v_P}{f}. \tag{7.26}$$

The positively propagating voltage wave is given by

$$V^+ = V_o^+ e^{-j\beta z} = |V_o^+| e^{j\phi} e^{-j\beta z}. \tag{7.27}$$

The time-domain form of this voltage is calculated from the phasor form as

$$v^+(t) = Re\left(|V_o^+| e^{j\phi} e^{-j\beta z} e^{j\omega t}\right)$$
$$= |V_o^+| \cos(\omega t - \beta z + \phi). \tag{7.28}$$

Most of the details of wave propagation that describe TEM plane waves are applicable for voltage and current waves as well. There is a corresponding current wave that accompanies the voltage wave. It satisfies the same wave equation and its solution is found in a similar manner. However, the current can be found directly from the scalar form of Faraday's law, Eq. (7.14), as

$$I^+(z) = \frac{\dfrac{\partial V^+(z)}{\partial z}}{-j\omega \mathcal{L}} = \frac{-j\beta}{-j\omega \mathcal{L}} V_o^+ e^{-j\beta z} = \frac{V_o^+ e^{-j\beta z}}{\dfrac{\omega \mathcal{L}}{\beta}}$$

$$= \frac{V_o^+ e^{-j\beta z}}{\dfrac{\omega \mathcal{L}}{\omega \sqrt{\mathcal{L}\mathcal{C}}}} = \frac{V_o^+ e^{-j\beta z}}{\sqrt{\dfrac{\mathcal{L}}{\mathcal{C}}}} = \frac{V_o^+ e^{-j\beta z}}{Z_C} \tag{7.29}$$

where

$$Z_o = \frac{V^+}{I^+} = \sqrt{\frac{\mathcal{L}}{\mathcal{C}}} \tag{7.30}$$

is defined as the *characteristic impedance* of the transmission line. This is the ratio of a positively propagating phasor voltage to the corresponding positively propagating phasor current. It is analogous to the wave impedance of TEM plane waves. A source connected to an infinite transmission line (so that there is only a positively propagating wave) will "see" an input impedance of Z_o at the two terminals of the line. Though this is not an actual impedance that can be localized at a specific location, it is a measurable quantity. The source would experience the same conditions if the infinite line were replaced by an impedance of value Z_o. Though it is a real number, Z_o does not represent power dissipation as with a resistor, but it does represent power taken from the source by the line. This power propagates through the line without any loss toward the other end of the line. Note that some books denote characteristic impedance by Z_C. For waves propagating in the $-z$-direction, the ratio of voltage to current is given by

$$\frac{V^-}{I^-} = -Z_o = -\sqrt{\frac{\mathscr{L}}{\mathscr{C}}}. \tag{7.31}$$

This does not signify a negative impedance. The sign change is due to the definition of positive current in the $+z$-direction while propagation is in the $-z$-direction analogous to the sign dependence of wave impedance with the direction of propagation.

Circuit methods are useful in calculating the power carried by the positively propagating voltage/current wave as

$$
\begin{aligned}
P_{AVE}^+ &= Re\left(\frac{V^+ I^{+*}}{2}\right) = Re\left(\frac{V_o^+ e^{-j\beta z} V_o^{+*} e^{+j\beta z}}{2 Z_o^*}\right) \\
&= \frac{|V_o|^2}{2 Z_o} = \frac{Z_o |V_o|^2}{2 Z_o^2} = \frac{Z_o I^2}{2} = \frac{I^2}{2 Y_o}
\end{aligned} \tag{7.32}
$$

where Z_o is real and $Y_o = 1/Z_o$ is called the *characteristic admittance*. We will use power calculations more in later sections.

The details of this section provide the essence of voltage and current waves on lossless transmission lines. Further features merely embellish or better describe the phenomena for special situations.

Example 7.4-1. Calculate the propagation constant and the characteristic impedance for the coaxial transmission line of Example 7.3-1. The inductance/meter is $\mathscr{L} = (\mu/2\pi)\ln(b/a)$; the capacitance/meter is $\mathscr{C} = 2\pi\varepsilon/\ln(b/a)$. The propagation constant is calculated as $\beta = \omega\sqrt{\mathscr{L}\mathscr{C}} = \omega\sqrt{\frac{\mu}{2\pi}\ln\left(\frac{b}{a}\right)\frac{2\pi\varepsilon}{\ln(b/a)}} = \omega\sqrt{\mu\varepsilon}$ as for plane waves. It should be of this form because the fields form a TEM wave. The characteristic impedance is calculated as $Z_o = \sqrt{\frac{\mathscr{L}}{\mathscr{C}}} = \sqrt{\frac{(\mu/2\pi)\ln(b/a)}{2\pi\varepsilon/\ln(b/a)}} = \sqrt{\frac{\mu}{\varepsilon}}\frac{\ln(b/a)}{2\pi} = \eta\frac{\ln(b/a)}{2\pi}$ which differs somewhat from TEM plane wave. The fields within the coaxial transmission line are quite different than those of a linearly polarized, TEM, plane wave.

7.5   LOSSY TRANSMISSION LINES

Actual transmission lines have metallic rather than PEC conductors with associated loss. In addition, dielectrics may have losses as well. These features can be added to our incremental circuit models for transmission lines very easily.

The high frequency behavior of current flow on the metallic conductors of transmission lines is accurately approximated by its existence only within the skin depth of the conductor.

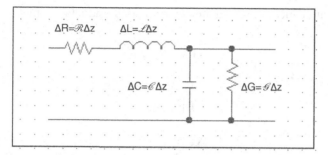

FIGURE 7.6: Incremental circuit model for lossy transmission line.

The resistance of this thin surface layer for a conductor of length L, cross-sectional area A, and perimeter P is calculated by

$$R = \frac{L}{\sigma A} = \frac{L}{\sigma \delta P}. \tag{7.33}$$

In order to use these results in the incremental circuit model, it is more useful in the form of resistance/meter that is defined as

$$\mathcal{R} = \frac{R}{L} = \frac{1}{\sigma A} = \frac{1}{\sigma \delta P}. \tag{7.34}$$

Current flow through this resistance causes a voltage drop that adds to that of the inductance as if the two elements are in series. This effect is readily included in the model with the addition of series incremental resistance as shown in Fig. 7.6. Since both conductors contribute to the voltage drop, the single incremental resistance must account for the series resistance of both conductors

$$\mathcal{R} = \frac{1}{\sigma \delta P_{\text{Total}}} = \frac{1}{\sigma \delta} \left(\frac{1}{P_1} + \frac{1}{P_2} \right). \tag{7.35}$$

Losses in the dielectric material of a transmission line are usually much less than the copper losses of the conductors. But, for completeness, they are included in the incremental circuit model of Fig. 7.6 as well. These dielectric losses result from an imaginary component of permittivity, $\varepsilon = \varepsilon' - j\varepsilon''$. Some books use an alternate description of an effective conductance within the dielectric, σ_{EFF},[*] to describe power dissipation. The dielectric power loss is calculated from Poynting's theorem and equated to similar calculations from circuit theory as

$$P_{\text{DIEL}} = \iiint\limits_{\forall} \frac{\sigma_{\text{EFF}} |\mathbf{E}|^2}{2} \, dv = \frac{V^2 G}{2} \tag{7.36}$$

---

[*] σ_{EFF} is not easily described by our models for dielectric material of Chapter 2. Since we don't often need it, we will defer its definition to more advanced courses.

from which the circuit conductance can be determined. Conductance/meter, $\mathscr{G}$, is calculated as

$$\mathscr{G} = \frac{\Delta G}{\Delta z} = \frac{\iiint\limits_\forall \sigma_{\text{EFF}}|\mathbf{E}|^2 dv}{V^2 \Delta z}$$

$$= \frac{\int\limits_{\Delta z} \iint\limits_{A_{\text{DIEL}}} \sigma_{\text{EFF}}|\mathbf{E}|^2 da\, dz}{V^2 \Delta z}$$

$$\approx \frac{\Delta z \iint\limits_{A_{\text{DIEL}}} \sigma_{\text{EFF}}|\mathbf{E}|^2 da}{V^2 \Delta z} = \frac{\iint\limits_{A_{\text{DIEL}}} \sigma_{\text{EFF}}|\mathbf{E}|^2 da}{V^2}. \qquad (7.37)$$

Fortunately, these losses are so small that $\mathscr{G} = 0$ is an accurate approximation for most transmission lines and we can omit this from most calculations.

The presence of this series voltage drop due to conductor resistance in the direction of the current flow is usually quite small for low-loss conductors. However, the total electric field is no longer strictly TEM throughout the transmission line. Near the conductors, there will be a small axial component of the electric field in the direction of current flow, see Fig. 7.7. The current and the axial component point in the $+z$-direction near one conductor; near the other conductor, they point in the $-z$-direction. Though the electric field in actual transmission lines is not exclusively transverse, the axial component is nearly negligible compared to the transverse component for a good transmission line. In cases where the axial component is significant, the line usually has too much loss to be useful.

The equations describing the behavior of a lossy line are obtained by KVL as

$$V(z) + (\Delta R + j\omega \Delta L)I(z) = V(z + \Delta z) \qquad (7.38)$$

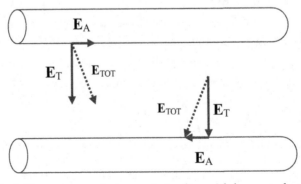

FIGURE 7.7: Electric field components in transmission line with lossy conductors.

which leads to the differential form

$$\frac{\partial V(z)}{\partial z} = -(\mathcal{R} + j\omega\mathcal{L})I(z) = -\mathcal{Z}I(z) \qquad (7.39)$$

and by KCL as

$$I(z) = (\Delta G + j\omega\Delta C)V(z) + I(z + \Delta z) \qquad (7.40)$$

which leads to the differential form

$$\frac{\partial I(z)}{\partial z} = -(\mathcal{G} + j\omega\mathcal{C})V(z) = -\mathcal{Y}V(z). \qquad (7.41)$$

For convenience, the series elements are combined as $\mathcal{Z} = \mathcal{R} + j\omega\mathcal{L}$ and $\mathcal{Y} = \mathcal{G} + j\omega\mathcal{C}$, impedance/meter and admittance/meter, respectively. The differential forms are combined to obtain

$$\frac{\partial^2 V(z)}{\partial z^2} = -\mathcal{Z}\frac{\partial I(z)}{\partial z} = \mathcal{Z}\mathcal{Y}V(z), \qquad (7.42)$$

a wave equation with the propagation constant of the form

$$\gamma = \pm\sqrt{\mathcal{Z}\mathcal{Y}} = \pm\sqrt{(\mathcal{R} + j\omega\mathcal{L})(\mathcal{G} + j\omega\mathcal{C})}. \qquad (7.43)$$

For lossless cases, $\gamma = j\beta$ as given by Eq. (7.24); for lossy cases, $\gamma = \alpha + j\beta$, an attenuated wave. In a similar fashion, the characteristic impedance becomes

$$Z_0 = \sqrt{\frac{\mathcal{Z}}{\mathcal{Y}}} = \sqrt{\frac{\mathcal{R} + j\omega\mathcal{L}}{\mathcal{G} + j\omega\mathcal{C}}}, \qquad (7.44)$$

a real quantity for lossless lines, but complex for lossy lines. Equations (7.43) and (7.44) are analogous to Eqs. (6.39) and (6.44) for TEM plane waves.

It is convenient to modify these equations for special cases. For the *lossless case*, these equations reduce to

$$\gamma = \pm j\beta = \pm j\omega\sqrt{\mathcal{L}\mathcal{C}} \qquad (7.45)$$

and

$$Z_0 = \frac{V^+}{I^+} = \sqrt{\frac{\mathcal{L}}{\mathcal{C}}}, \qquad (7.46)$$

respectively, resulting in unattenuated waves propagating on the line.

For *low-loss cases*, $\mathcal{R} \ll \omega\mathcal{L}$ and $\mathcal{G} \ll \omega\mathcal{C}$, the propagation constant becomes

$$\gamma = j\omega\sqrt{\mathcal{L}\mathcal{C}}\sqrt{\left(1+\frac{\mathcal{R}}{j\omega\mathcal{L}}\right)\left(1+\frac{\mathcal{G}}{j\omega\mathcal{C}}\right)}$$

$$\approx j\omega\sqrt{\mathcal{L}\mathcal{C}}\left(1+\frac{\mathcal{R}}{j2\omega\mathcal{L}}\right)\left(1+\frac{\mathcal{G}}{j2\omega\mathcal{C}}\right)$$

$$\approx j\omega\sqrt{\mathcal{L}\mathcal{C}}\left(1-\frac{j}{2}\left(\frac{\mathcal{R}}{\omega\mathcal{L}}+\frac{\mathcal{G}}{\omega\mathcal{C}}\right)\right)$$

$$= \left(\frac{\mathcal{R}}{2Z_C}+\frac{\mathcal{G}}{2Y_C}\right)+j\omega\sqrt{\mathcal{L}\mathcal{C}} = \alpha + j\beta. \qquad (7.47)$$

The wave experiences attenuation due to the copper and dielectric losses and phase shift due to the stored electric and magnetic energy. Z_o, the ratio of positively propagating phasor voltage to positively propagating phasor current, is approximated by

$$Z_o = \sqrt{\frac{\mathcal{L}}{\mathcal{C}}} = \sqrt{\frac{\mathcal{R}+j\omega\mathcal{L}}{\mathcal{G}+j\omega\mathcal{C}}} = \sqrt{\frac{\mathcal{L}}{\mathcal{C}}}\sqrt{\frac{\left(1+\frac{\mathcal{R}}{j\omega\mathcal{L}}\right)}{\left(1+\frac{\mathcal{G}}{j\omega\mathcal{C}}\right)}}$$

$$\approx \sqrt{\frac{\mathcal{L}}{\mathcal{C}}}\sqrt{\left(1+\frac{\mathcal{R}}{j\omega\mathcal{L}}\right)\left(1-\frac{\mathcal{G}}{j\omega\mathcal{C}}\right)}$$

$$\approx \sqrt{\frac{\mathcal{L}}{\mathcal{C}}}\left(1-\frac{j}{2}\left(\frac{\mathcal{R}}{\omega\mathcal{L}}-\frac{\mathcal{G}}{\omega\mathcal{C}}\right)\right), \qquad (7.48)$$

a complex characteristic impedance for lossy transmission lines.

The vast majority of actual transmission lines have significant copper losses, but negligible dielectric losses. For this case, the propagation constant becomes

$$\gamma = \frac{\mathcal{R}}{2Z_o}+j\omega\sqrt{\mathcal{L}\mathcal{C}}. \qquad (7.49)$$

Even though the real part of the propagation constant is very small, it cannot be neglected. It is multiplied by the distance term z and this product must be retained since it occurs in the exponent and for large distances has a significant attenuation effect, $e^{-\alpha z}$. On the other hand, the imaginary part of the characteristic impedance does not show this effect with distance; when it is small, its effects are small regardless of the line length. Consequently, the characteristic impedance is approximated as real for all but the most exacting applications,

$$Z_o \approx \sqrt{\frac{\mathcal{L}}{\mathcal{C}}}. \qquad (7.50)$$

Note that a few books emphasize the resistive nature of Z_o by calling it *characteristic resistance* and denoting it by R_o.

No further cases need to be considered since a highly lossy transmission line would attenuate a signal too much to be useful.

Example 7.5-1. A transmission line has distributed parameters of $\mathscr{L} = 5$ μH/m, $\mathscr{C} = 0.02$ μF/m, and $\mathscr{R} = 0.01$ Ω/m. Calculate γ and Z_o for operation at $\omega = 10^6$ rad/s. From Eq. (7.43), $\gamma = \sqrt{ZY} = \sqrt{(0.01 + j10^6(5 \times 10^{-6}))\, j10^6(0.02 \times 10^{-6})} = 0.000316 + j0.316$ m^{-1}.

From Eq. (7.44), $Z_o = \sqrt{\dfrac{\mathscr{Z}}{\mathscr{Y}}} = \sqrt{\dfrac{0.01 + j10^6(5\times10^{-6})}{j10^6(0.02\times10^{-6})}} = 15.8 - j0.0158$ Ω.

This appears to be a low-loss case since $\mathscr{R}/\omega\mathscr{L} = 0.01/5 = 0.002 \ll 1$.

Example 7.5-2. Obtain expressions for γ and Z_o for the unusual looking incremental circuit model shown in Fig. 7.8.

From Eq. (7.43), $\gamma = \sqrt{\mathscr{Z}\mathscr{Y}} = \sqrt{j\omega\mathscr{R}\mathscr{C}} = \dfrac{1+j}{\sqrt{2}}\sqrt{\omega\mathscr{R}\mathscr{C}}$, and from Eq. (7.44), $Z_o = \sqrt{\dfrac{\mathscr{Z}}{\mathscr{Y}}} = \sqrt{\dfrac{\mathscr{R}}{j\omega\mathscr{C}}} = \dfrac{1-j}{\sqrt{2}}\sqrt{\dfrac{\mathscr{R}}{\omega\mathscr{C}}}$.

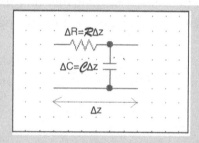

FIGURE 7.8: Incremental model for a line without inductance.

Example 7.5-3. Calculate the ratio of $E_{\text{AXIAL}}/E_{\text{TRANS}}$ for a typical coaxial line with copper conductors of radii of 0.0005 m and 0.005 m, respectively, with a 10 V potential difference between conductors. The line has 50 Ω impedance and operates at 10^6 Hz. From Eq. (7.6), $|E_{\text{TRAN}}|_{\text{MAX}} = V_o/[\ln(b/a)\rho]_{\text{INNER}} = 10/[\ln(10)(0.0005)] = 8686$ V/m. E_{AXIAL}: near the inner conductor is due to current flow in the skin depth of the inner conductor as $|E_{\text{AXIAL}}|_{\text{MAX}} = I\mathscr{R} = (10/50)/(2\pi a\sigma\delta) = (0.1/\pi 5.8 \times 10^7)(10^3/0.067)(1/0.0005) = 0.0164$ V/m. Therefore, $E_{\text{AXIAL}}/E_{\text{TRAN}} = 0.0164/8686 = 1.89 \times 10^{-6}$, very close to TEM!

7.6 REFLECTED WAVES

Reflection of propagating TEM plane waves occurs whenever the wave encounters a material boundary. A similar phenomenon occurs for voltage and current waves on transmission lines whenever they are incident on a boundary between different transmission lines. The details of this phenomenon for the transmission line connection are shown in Fig. 7.9. For simplicity,

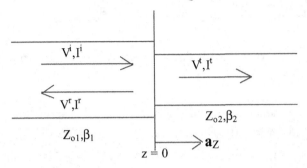

FIGURE 7.9: Transmission line junction.

both lines are assumed lossless. The left-hand line for $z < 0$ is represented by β_1 and Z_{C1}; the right-hand line for $z > 0$ as β_2 and Z_{C2}. The different spacing of the symbolic conductors for the two lines is to show that they have different characteristic impedances and propagation constants. A voltage wave from the right,

$$V^i = V_o e^{-j\beta_1 z} \tag{7.51}$$

is incident on the boundary; the accompanying current wave is given by

$$I^i = \frac{V_o}{Z_{o1}} e^{-j\beta_1 z}. \tag{7.52}$$

A reflected wave, proportional to the incident wave, propagates away from the boundary in the $-z$-direction as

$$V^r = \Gamma V_o e^{+j\beta_1 z} \tag{7.53}$$

where the reflection coefficient, Γ, is defined as

$$\Gamma = \left. \frac{V^r}{V^i} \right|_{\text{BOUNDARY}}; \tag{7.54}$$

the reflected current wave is given by

$$I^r = -\frac{\Gamma V_o}{Z_{o1}} e^{+j\beta_1 z}. \tag{7.55}$$

Similarly, a transmitted wave propagates away from the boundary in the $+z$-direction as

$$V^t = T V_o e^{-j\beta_2 z} \tag{7.56}$$

where the transmission coefficient, T, is defined as

$$T = \left. \frac{V^t}{V^i} \right|_{\text{BOUNDARY}}. \tag{7.57}$$

The transmitted current wave is given by

$$I^t = \frac{TV_o}{Z_{o2}} e^{-j\beta_2 z}. \tag{7.58}$$

The "boundary conditions" for voltages and currents are merely KVL and KCL. The voltages on both sides of the boundary must be equal as

$$(V^i + V^r)|_{z=0} = V_0(1 + \Gamma) = V^t|_{z=0} = TV_o. \tag{7.59}$$

Likewise the currents entering one side of the boundary must equal the currents leaving the other

$$(I^i + I^r)|_{z=0} = V_0 \left(\frac{1 - \Gamma}{Z_{o1}} \right) = I^t|_{z=0} = \frac{TV_o}{Z_{o2}}. \tag{7.60}$$

The solution to the two equations is

$$\Gamma = \frac{Z_{o2} - Z_{o1}}{Z_{o2} + Z_{o1}} \tag{7.61}$$

and

$$T = \frac{2Z_{o2}}{Z_{o2} + Z_{o1}}, \tag{7.62}$$

forms similar to the reflection and transmission coefficients of TEM plane waves. Though derived for a semi-infinite second transmission line, Z_{o2}, the second line could be replaced by a lumped impedance with the same results. The signal incident upon the boundary at $z = 0$ where the load terminals are attached would not see any difference in the voltage and current of the lumped impedance since there is no wave in line 2 propagating toward the boundary from the right. We will deal later with the more complicated situation of waves in both directions on a finite length for line 2. When the reflections occur from a lumped impedance, the load reflection coefficient is often written with a subscript as Γ_L.

To eliminate reflections, it is a common practice to terminate a line with its characteristic impedance, $Z_L = Z_{o2} = Z_{o1}$. This is known as *matching the line* or using a *matched load* and results in $\Gamma_{\text{MATCH}} = 0$ with no reflected wave. A short circuit, $Z_L = 0$, produces $\Gamma_{\text{SC}} = -1$. The incident voltage must be canceled by an equal, but opposite reflected, voltage in order that there is no voltage drop across the short circuit. As can be seen from Eq. (7.60), the current in a short circuit will be double the incident current. An open circuit, $Z_L = \infty$, produces $\Gamma_{\text{OC}} = +1$. The total current must be zero at an open circuit so that the reflected current is equal, but oppositely directed to the incident current. Consequently, the voltage at an open circuit is double the incident voltage. Reactive loads, $Z_L = jX$, produce reflection coefficients with $\Gamma_L = 1e^{j\phi}$; the reflection coefficient is complex, of unit magnitude, and with the phase angle dependent upon the value and sign of the reactive load, jX. These several examples illustrate an important principle: *the reflection coefficient for a passive load is a complex number with a magnitude of 1 or less.* In reality, it is impossible to fabricate a perfect short or open or a lossless reactance. Consequently, actual loads always have reflection coefficients with a magnitude somewhat less than unity.

The current of a wave is related to the wave's voltage by the characteristic impedance and so the reflected currents are related to the reflected voltages. But, since the voltage–current ratios for oppositely propagating waves are of opposite sign, $V^+/I^+ = Z_o = -V/I^-$, the current reflects with an opposite sign of the reflected voltage. Some books and literature emphasize this point by separately defining both a voltage and a current reflection coefficient as

$$\Gamma_I = \left. \frac{I^r}{I^i} \right|_{\text{LOAD}} = -\Gamma_V = -\frac{Z_2 - Z_o}{Z_2 + Z_o}. \tag{7.63}$$

For a short circuit, the reflected voltage must reverse the sign of the incident voltage in order to keep the load voltage zero with $\Gamma_V = -1$; the corresponding current reflection coefficient is $\Gamma_I = +1$. For an open circuit, the reflected current must reverse sign so that the load current is zero with $\Gamma_I = -1$. Most literature doesn't explicitly use the current reflection coefficient, but uses this sign-reversal concept for reflected currents.

These principles of reflected waves enable us to look more closely at waves on transmission lines. We will first examine transient waves followed by steady-state waves.

7.7   TRANSIENT WAVES ON LINES

The incremental circuit model is the basis of the transient wave analysis. For general transient analysis, we express the voltages, currents, and impedances in the Laplace s-domain; inversion of the s-domain solution gives the time-domain result. The incremental model in the s-domain is shown in Fig. 7.10. Application of KVL and KCL to this model leads to the s-domain PDEs as

$$\frac{\partial V(z, s)}{\partial z} = -\boldsymbol{\mathcal{Z}}(s)I(z, s) = -[\boldsymbol{\mathcal{R}}(s) + s\boldsymbol{\mathcal{L}}(s)]I(z, s)$$

$$\frac{\partial I(z, s)}{\partial z} = -\boldsymbol{\mathcal{Y}}(s)V(z, s) = -[\boldsymbol{\mathcal{G}}(s) + s\boldsymbol{\mathcal{C}}(s)]V(z, s) \tag{7.64}$$

which can be combined to give a wave equation

$$\frac{\partial^2 V(z, s)}{\partial z^2} = \boldsymbol{\mathcal{Z}}(s)\boldsymbol{\mathcal{Y}}(s)V(z, s) \tag{7.65}$$

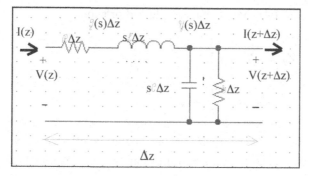

FIGURE 7.10: The s-domain incremental circuit model for a transmission line.

and leads to a solution of

$$V(z, s) = V^+(s)e^{-\sqrt{\mathcal{3}\mathcal{4}}z} + V^-(s)e^{\sqrt{\mathcal{3}\mathcal{4}}z}. \tag{7.66}$$

Since, the time-domain solution is of great interest for transients, the inverse transform of $V(z, s)$ is needed. For lossless lines, $\sqrt{\mathcal{3}\mathcal{4}} = s\sqrt{\mathcal{L}\mathcal{C}}$ and the s-domain voltage becomes

$$V(z, s) = V^+(s)e^{-\sqrt{\mathcal{L}\mathcal{C}}sz} + V^-(s)e^{\sqrt{\mathcal{L}\mathcal{C}}sz}. \tag{7.67}$$

As you recall, s-domain functions of the form $F(s)e^{-as}$ has a time-domain solution of the form $f(t - a)$ where $f(t)$ is the inverse transform of $F(s)$. Therefore, the time-domain solution for the voltage is expressed as

$$v(z, t) = v^+(t - \sqrt{\mathcal{L}\mathcal{C}}z) + v^-(t + \sqrt{\mathcal{L}\mathcal{C}}z) \tag{7.68}$$

where $v^+(t - \sqrt{\mathcal{L}\mathcal{C}}z)$ and $v^-(t + \sqrt{\mathcal{L}\mathcal{C}}z)$ represent voltage waves which are moving in the $+z$- and $-z$-directions, respectively. An observer at $z = 0$ would see them as $v^+(t)$ and $v^-(t)$. However, an observer at z will see the signal delayed by $t_d = \pm\sqrt{\mathcal{L}\mathcal{C}}z$; the delay of the signal is due to the finite propagation time of waves on the line. From this relationship, we see that $v_P = \pm 1/\sqrt{\mathcal{L}\mathcal{C}}$. Therefore, the magnitude and waveshape of a signal at the generator terminals is preserved as it travels down the line in either direction with velocity v_P. This is consistent with our findings for steady-state waves.

The problem is somewhat more complicated for typical low-loss lines. The impedance/meter is expressed as

$$\mathcal{3} = \mathcal{R} + s\mathcal{L} = \frac{\sqrt{s}}{P}\sqrt{\frac{\mu}{\sigma}} + s\mathcal{L} \tag{7.69}$$

where P is the perimeter of a metallic conductors. The skin depth frequency dependence in the series resistance/meter complicates the form. Note that the first term includes the internal inductance part of Eq. (6.64) that we have neglected previously,

$$Z_w = R_w + jX_w = \frac{\gamma}{\sigma} = \frac{1+j}{\sigma\delta} = \frac{1+j}{\sigma}\sqrt{\frac{\omega\mu\sigma}{2}}$$

$$= \sqrt{j\omega}\sqrt{\frac{\mu}{\sigma}} = \sqrt{s}\sqrt{\frac{\mu}{\sigma}}. \tag{7.70}$$

The additional inductance is very small and has negligible effect on the impedance/meter, $\mathcal{L}$. The "square" subscript is used to denote "ohms/square." The exponential term in the solution becomes

$$\sqrt{\mathcal{ZY}}z = \sqrt{(\mathcal{R} + s\mathcal{L})\,s\mathcal{C}z}$$

$$= \sqrt{\left(\frac{\sqrt{s}}{P}\sqrt{\frac{\mu_{\mathrm{CU}}}{\sigma_{\mathrm{CU}}}} + s\mathcal{L}\right)s\mathcal{C}z}$$

$$= s\sqrt{\mathcal{L}\mathcal{C}}\sqrt{\left(1 + \frac{1}{\sqrt{s}\,P\mathcal{C}}\sqrt{\frac{\mu_{\mathrm{CU}}}{\sigma_{\mathrm{CU}}}}\right)z}$$

$$\approx s\sqrt{\mathcal{L}\mathcal{C}}z + \sqrt{s}\frac{Z_0}{P}\sqrt{\frac{\mu_{\mathrm{CU}}}{\sigma_{\mathrm{CU}}}}z$$

$$= \frac{s\,z}{v_P} + \sqrt{s}\frac{Z_0}{P}\sqrt{\frac{\mu_{\mathrm{CU}}}{\sigma_{\mathrm{CU}}}}z \tag{7.71}$$

where the "CU" subscripts denote properties of the copper conductors. The first term in the exponential represents the delay due to finite propagation velocity that we have already discussed. The second term, due to the skin depth variation with frequency, complicates the inverse Laplace transform. Nevertheless, a solution can be obtained for the special case of the step function at the input which establishes only a positively propagating wave, i.e., $V^+(s) = 1/s$ and $V^-(s) = 0$, which leads to

$$V(z,s) = \frac{e^{-\sqrt{s}\frac{Z_0}{P}\sqrt{\frac{\mu_{\mathrm{CU}}}{\sigma_{\mathrm{CU}}}}z}}{s}e^{-s\frac{z}{v_P}} = V(s)e^{-s\frac{z}{v_P}}. \tag{7.72}$$

This form describes a time-domain delayed function $v(t - z/v_P)$ as the inverse Laplace transform of $V(s)$. Though not a well-known transform, MAPLE recognizes it and gives the solution of

$$v(t) = \mathbf{L}^{-1}\left\{\frac{e^{-\sqrt{s}\frac{Z_0}{P}\sqrt{\frac{\mu_{\mathrm{CU}}}{\sigma_{\mathrm{CU}}}}z}}{s}\right\} = \mathrm{erfc}\left(\frac{Z_0\sqrt{\frac{\mu_{\mathrm{CU}}}{\sigma_{\mathrm{CU}}}}z}{2P\sqrt{t}}\right) \tag{7.73}$$

where $\mathbf{L}^{-1}$ represents the inverse Laplace transform. $\mathrm{erfc}(x)$ is the complementary error function defined as

$$\mathrm{erfc}(x) = 1 - \frac{2}{\sqrt{\pi}} \int_0^x e^{-u^2}\,du. \tag{7.74}$$

Applying Eq. (7.72) to these results, we obtain the solution

$$v(z, t) = \mathrm{erfc}\left(\frac{Z_0 \sqrt{\frac{\mu_{CU}}{\sigma_{CU}}}\, z}{2P\sqrt{t - \frac{z}{v_P}}}\right). \tag{7.75}$$

The time delay accounts for the propagation of the discontinuity down the line. The results for a typical case of a 50 Ω, coaxial transmission line with $v_P = 10^8$ m/s and an inner copper conductor of radius of 0.0005 m and an outer copper conductor of radius of 0.005 m has been plotted in Fig. 7.11 for distances of 100 m and 500 m, respectively.

The skin effect losses produce greater loss for the higher frequencies within the step function applied to the input to the line. They also introduce *dispersion*, i.e., different frequencies propagate with different velocities. The wave diffuses down the line, spreading out as it propagates. This results in a greatly distorted version of the step with significantly increased time for the wave to reach 90% of final value. The response to the unit step on a comparable lossless line are shown for comparison. Fortunately, this advanced mathematics is seldom needed except for demanding applications where rise times are critical.

The complexity of the low-loss solution masks the essential features of the lossless solution. From here onward, we consider only the lossless solution characterized by magnitude and waveshape preservation and time delay due to finite velocity of propagation.

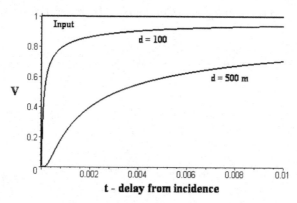

FIGURE 7.11: Response to a unit step of a coaxial line with skin effect losses.

7.8 MORE ABOUT TRANSIENT WAVES

The initiation of waves on a transmission line by the application of a generator is a common application. How do generators, transmission lines, and load impedances interact? Consider the signal generator with a Thévenin impedance $Z_S = R_S$ and an open circuit voltage of $u(t)$ attached to a line with characteristic impedance Z_o, velocity of propagation v_P, and no initial voltages or current, see Fig. 7.12. The line of length d has a load impedance Z_L attached to the far end. What is the response of the system?

A qualitative description of the response is that a wave, initiated at the input of the line by the generator, travels down the line with finite velocity until it reaches the far end. If the load impedance is not equal to the characteristic impedance, a reflected wave occurs which travels back to the source. When it reaches the source, a second reflection may occur and the process repeats itself. Now let's quantitatively describe the events.

Transient analysis of transmission lines is most easily accomplished by "following" the wave discontinuities as they propagate throughout the system. For the system of Fig. 7.12, there are no initial voltages or currents on the line; the voltage and current discontinuities occur at $t = 0$ for the unit step of the signal generator. This action results in a transient voltage wave that propagates down the line toward the load. The amplitude of this transient wave is determined in the following manner. The total voltage on a transmission line is the sum of two oppositely propagating waves. But, at the instant of the unit step discontinuity, the only wave possible is from the source toward the load; therefore, the total voltage at the line input is the positively propagating wave. The input impedance to the line is the ratio of total voltage to total current and is the same as that of the wave propagating toward the load, Z_o. Sometimes an alternate name of *surge impedance* is used to describe the transient impedance of a transmission line. At this instant and until a reflected wave appears at the line input, the line can be replaced by a lumped impedance equal to Z_o, see Fig. 7.13.

Since the generator "sees" a load of Z_o, its open-circuit voltage produces a voltage across the line input according to the voltage divider equation. In general, this expression is written in

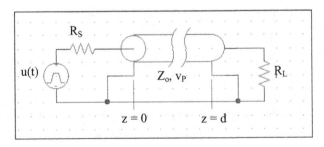

FIGURE 7.12: Pulse generator–transmission line system.

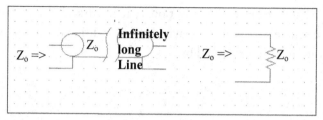

FIGURE 7.13: Transmission line and its transient equivalent input impedance.

the s-domain so that impedances that are functions of s can be handled. However, when the impedances are resistive, the time-domain form can be used directly

$$v_{IN}(t) = v^+{}_0(t) = v_{OC}(t)\frac{Z_o}{R_S + Z_o} = u(t)\frac{Z_o}{R_S + Z_o}. \tag{7.76}$$

The subscript 0 on v^+ indicates that this wave, the first of many waves propagating toward the load, has undergone no reflections. As the voltage discontinuity propagates down the line, the discontinuity in voltage moves with velocity v_P; its position is described by $z = v_P t$. The space–time description of the wave is expressed by

$$v_0^+(z, t) = u\left(t - \frac{z}{v_P}\right)\frac{Z_o}{R_S + Z_o}. \tag{7.77}$$

The time that the unit step occurs at position z is delayed due to the finite propagation velocity of the line. The discontinuity reaches the load at the end of the line at time

$$t_d = \frac{d}{v_P}. \tag{7.78}$$

The *delay time* of the line is an alternate characterization of the line that includes both the line length and propagation velocity. If the load is not matched to the line, a reflection described by

$$\Gamma_L = \frac{Z_L - Z_o}{Z_L + Z_o} \tag{7.79}$$

occurs at the load. For general impedances, the *load reflection coefficient*, Γ_L, is complex, but for resistive loads only, it is real and time-domain calculations can be used. For a resistive load, $Z_L = R_L$, the first of many possible reflected waves at the load is given by

$$v_1^-(d, t) = \Gamma_L v_1^+(d, t)$$

$$= u\left(t - \frac{d}{v_P}\right)\left(\frac{R_L - Z_o}{R_L + Z_o}\right)\frac{Z_o}{R_S + Z_o}. \tag{7.80}$$

When this reflected wave reaches position z, it has traveled a distance $d-z$ from the load back toward the source so it will be delayed by this additional amount. Consequently, the reflected wave at position z is represented by

$$v_1^-(z, t) = u\left(t - \frac{d}{v_P} - \frac{d-z}{v_P}\right)\left(\frac{R_L - Z_o}{R_L + Z_o}\right)\frac{Z_o}{R_S + Z_o}$$

$$= u\left(t - \frac{2d}{v_P} + \frac{z}{v_P}\right)\left(\frac{R_L - Z_o}{R_L + Z_o}\right)\frac{Z_o}{R_S + Z_o}. \qquad (7.81)$$

Note that the mathematics has included automatically a change of sign of the z/v_P term to indicate that the reflected wave is traveling in the opposite direction on the line.

When the reflected wave reaches the source, it may be reflected by the source. As before, the reflection coefficient depends upon the impedance "seen" by the wave approaching a discontinuity. The impedance of the source is its Thévenin impedance, R_S, so the *source reflection coefficient* is given by

$$\Gamma_S = \frac{Z_S - Z_o}{Z_S + Z_o} = \frac{R_S - Z_o}{R_S + Z_o}. \qquad (7.82)$$

This leads to a wave reflected from the source as

$$v_1^+(0, t) = \Gamma_S v_0^-(0, t)$$

$$= u\left(t - \frac{2d}{v_P}\right)\left(\frac{R_S - Z_o}{R_S + Z_o}\right)\left(\frac{R_L - Z_o}{R_L + Z_o}\right)\frac{Z_o}{R_S + Z_o}$$

$$= u\left(t - \frac{2d}{v_P}\right)\frac{Z_o}{R_S + Z_o}\Gamma_S\Gamma_L. \qquad (7.83)$$

The original signal has made one round trip down the line and back with a corresponding delay of $2t_d = 2d/v_P$. The amplitude has been altered by the product of the two reflection coefficients. This wave is now beginning the propagation down the line exactly as is done by the original wave from the source. We could complete another round of calculations, but with a little thought, we see that each round trip modifies the original wave by adding a delay of $2t_d$ and altering the magnitude by an additional multiplication by $\Gamma_S\Gamma_L$. Furthermore, these processes continue indefinitely resulting in an infinite series of waves on the line. Each successive round trip reduces the wave amplitude and delays it further and further. A careful consideration of these terms leads to

$$v(z, t) = \sum_{n=0}^{N} v_n^{+}(z, t) + \sum_{n=1}^{N} v_n^{-}(z, t)$$

$$= \sum_{n=0}^{N} u\left(t - 2nt_d - \frac{z}{v_P}\right) \frac{Z_o}{R_S + Z_o} (\Gamma_S \Gamma_L)^n$$

$$+ \sum_{n=0}^{N} u\left(t - 2(n+1)t_d + \frac{z}{v_P}\right) \frac{Z_o \Gamma_L}{R_S + Z_o} (\Gamma_S \Gamma_L)^n$$

$$= \frac{Z_o}{R_S + Z_o} \sum_{n=0}^{N} \left\{ \begin{array}{l} u\left(t - 2nt_d - \frac{z}{v_P}\right) \\ +u\left(t - 2(n+1)t_d + \frac{z}{v_P}\right)\Gamma_L \end{array} \right\} (\Gamma_S \Gamma_L)^n \qquad (7.84)$$

where N is determined by the time of interest according to $Nt_d < t < (N+1)t_d$. Note that when either the load or the source is matched, i.e., the load or the source reflection coefficient is zero, then the infinite series has only one or two terms, respectively. Each term in the series represents the passage of a wavefront; the amplitude and time of passage is determined by the location on the line and the time since the wave began. The passage of each wavefront introduces a discontinuity to the existing voltage so that the total voltage is the previous value plus the voltage of the discontinuity. In other words, the total voltage at any instant is the original voltage (in this case zero) plus the sum of all the wavefronts that have passed. After many reflections, the wavefronts are of such small amplitude that the system can be said to be in steady state. This can be determined by the summation of the infinite series as

$$v(z, \infty) = \frac{Z_o}{R_S + Z_o} \sum_{n=0}^{N} (1 + \Gamma_L)(\Gamma_S \Gamma_L)^n$$

$$= \frac{Z_o (1 + \Gamma_L)}{R_S + Z_o} \sum_{n=0}^{N} (\Gamma_S \Gamma_L)^n$$

$$= \frac{Z_o (1 + \Gamma_L)}{(R_S + Z_o)(1 - \Gamma_S \Gamma_L)}. \qquad (7.85)$$

There is no longer any time dependence because all of the transient wavefronts have passed and all of the unit step functions are "on". For physical source and load impedances, $\Gamma_S \Gamma_L < 1$ and the geometric series in $\Gamma_S \Gamma_L$ is given in closed form as $\sum a^n = 1/(1 - a)$. This can be expressed in terms of the source and load impedances as

$$
\begin{aligned}
v(z, \infty) &= \frac{Z_o \left(1 + \Gamma_L\right)}{(R_S + Z_o)\left(1 - \Gamma_S \Gamma_L\right)} \\
&= \frac{Z_o \left(1 + \frac{R_L - Z_o}{R_L + Z_o}\right)}{(R_S + Z_o)\left(1 - \left(\frac{R_S - Z_o}{R_S + Z_o}\right)\left(\frac{R_L - Z_o}{R_L + Z_o}\right)\right)} \\
&= \frac{Z_o \left(\frac{2 R_L}{R_L + Z_o}\right)}{(R_S + Z_o)\left\{\frac{2 Z_o (R_S + R_L)}{(R_S + Z_o)(R_L + Z_o)}\right\}} = \frac{R_L}{R_S + R_L}.
\end{aligned}
\tag{7.86}
$$

This is the DC value of the load voltage when the load is directly connected to the source! There is no dependence upon z because all points on the line are at the same potential; all the transients have decayed to zero. On the other hand, a lossy line will affect the steady-state voltages since it introduces a resistance between the source and the load due to copper losses in the line.

An implementation of Eq. (7.87) is shown in Fig. 7.14 for the parametric values of $Z_C = 50 \ \Omega$, $R_S = 25 \ \Omega$, $R_L = 100 \ \Omega$, $v_P = 10^8$ m/s, and $t_d = 1 \ \mu$s. Note that only $N = 5$ terms are needed to obtain a plot of the system response to a unit step input voltage. The voltages at the input, the midpoint and the load are all shown. The results satisfy Eqs. (7.76) and (7.86) at $t = 0$ and $t = \infty$, respectively.

An alternate, graphical procedure, known as a bounce diagram, is quick, easy, and it gives a physical feeling to the process. In addition, it is especially useful for more complicated situations involving several, interconnected transmission lines. This is the subject of the next section.

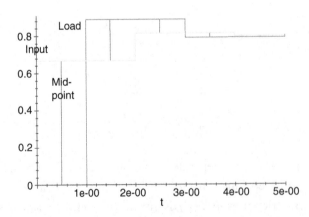

FIGURE 7.14: Transient response of transmission line system to unit step.

7.9   BOUNCE DIAGRAMS

The process of accounting for the multiple wavefronts can be connected with the physical model of propagation delays and reflections by a simple graphical process. The method is best explained by considering again the system of Fig. 7.12.

The generation of the unit step (or in general, any discontinuity) by the generator initiates a wavefront at the left end of the line that propagates toward the load. The amplitude of the wave is calculated from the circuit model of Fig. 7.12 according to

$$v_{IN}(t) = v_0^+(t) = v_{OC}(t)\frac{Z_o}{R_S + Z_o} = u(t)\frac{Z_o}{R_S + Z_o}. \tag{7.87}$$

Since the wave propagates with velocity v_P, the progress of the wavefront in space and time can be plotted as a line of constant slope, $v_P = \Delta z/\Delta t$, reaching the load at time $t = t_d = d/v_P$, see trace A of Fig. 7.15. The amplitude of the wavefront is indicated adjacent to the trace. Upon reaching the load, the wavefront is reflected; its amplitude is multiplied by the load reflection coefficient and a wave in the opposite direction is initiated. The new wave propagates with the same constant slope as the previous wave, but since it is moving in the opposite direction, it has an opposite slope, see trace B of Fig. 7.15. This wave reaches the source at time $t = 2t_d$ and the reflection (by the source reflection coefficient) initiates a second wavefront in the forward direction, see trace C of Fig. 7.15. This process continues indefinitely with each the magnitude of each wavefront multiplied by an additional reflection coefficient, Γ_S or Γ_L, and delayed by an additional time delay, t_d. The observed voltage at position z_o and time t_o is the summation of all previous wavefronts that have passed this point for $t \le t_o$. The waveshape at position z_o as a function of time is constructed as the sequential summation of the wavefronts which

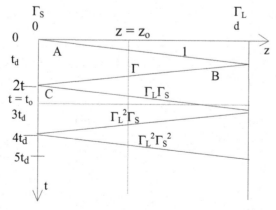

FIGURE 7.15: Bounce diagram.

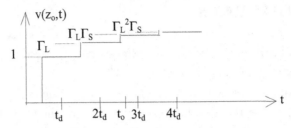

FIGURE 7.16: Voltage observed at line location z_0 as a function of time from bounce diagram, Fig. 7.15.

intersect a vertical line at $z = z_0$, see Fig. 7.16. The voltage distribution along the line for time t_0 can be found as the summation of wavefronts which have preceded the horizontal line $t = t_0$, see Fig. 7.17. Corresponding currents associated with each wavefront are easily included as well.

The response to a unit step input provides the basis for analyzing the response to any waveshape via step-wise approximation of the exciting signal and superposition of the responses. Pulses can be analyzed as well since they are merely two steps—one positive, the other negative—with an intervening delay equal to the pulse width. The effects of pulse width compared to line delay greatly affect the total response to multi-step signals.

This method can be extended to include several sections of line, also. The incident voltage at the junction between two lines establishes both a reflected and a transmitted wave. The transmitted wave propagating into the second line can be plotted as an incident wave in the second line. If it undergoes reflection at some load, the reflected wave will impinge on the junction and establish a wave in the first line propagating back to the source. These steps become rather tedious as more and more waves must be considered. But the principles remain simple, easy to apply, and revealing of the physical processes.

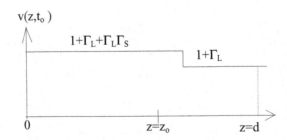

FIGURE 7.17: Voltage distribution on line at time t_0.

Example 7.9-1. Using the bounce diagram method, determine the voltage at the input to the line of Fig. 7.12 for $Z_o = 50\ \Omega$, $R_S = 25\ \Omega$, $R_L = 100\ \Omega$, $d = 100$ m, and $v_P = 10^8$ m/s. From these parameters, we obtain $t_d = 1$ μs, $\Gamma_S = -1/3$, and $\Gamma_L = 1/3$. The bounce diagram is shown in Fig. 7.18(a); the input voltage is shown in Fig. 7.18(b) and compares agrees with the computer-generated solution shown in Fig. 7.14.

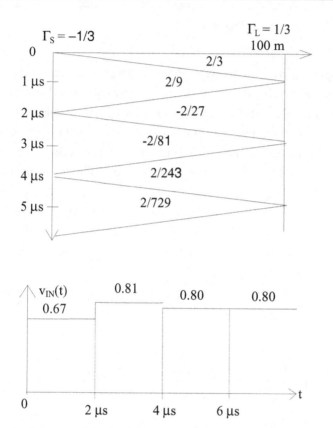

FIGURE 7.18: Bounce diagram method: (a) bounce diagram and (b) input voltage.

Example 7.9-2. Using the bounce diagram method, determine the voltage at the junction of the three equal-length transmission lines shown in Fig. 7.19. Since, the line on the lower right is terminated in its characteristic impedance, it can be replaced by a lumped 50 Ω resistor at the junction. A wave approaching the junction from the 50 Ω line on the left "sees" $50||25 = 16^2/3\ \Omega$ which gives a reflection coefficient of $\Gamma_{\text{LEFT}} = -1/2$. A wave approaching the junction from the 25 Ω line on the right "sees" $50||50 = 25\Omega$ which gives a reflection coefficient of $\Gamma_{\text{RIGHT}} = 0$. Assuming that all three lines have the same phase

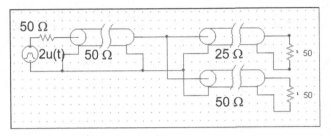

FIGURE 7.19: Multi-transmission line junction.

velocity of propagation, we can sketch the bounce diagram as shown in Fig. 7.20(a). The initial wave down the 50 Ω line is partially reflected back toward the source and partially transmitted down the 50 Ω and 25 Ω lines on the right. No reflection occurs from load on the 50 Ω line. However, a reflection occurs on the 25 Ω line and propagates back to the junction where it undergoes no reflection, but total transmission toward the source on the left-most 50 Ω line. There are no further reflections. The junction voltage is shown in Fig. 7.20(b).

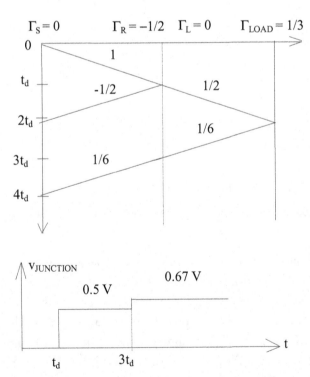

FIGURE 7.20: Bounce diagram method for several lines: (a) bounce diagram and (b) junction voltage.

7.10 REACTIVE LOAD TRANSIENTS

Reflection of transient waves from capacitors, inductors, or general reactive loads are more complicated due to the frequency dependence of their impedance. The general principles of these reflections can be seen by considering a unit step incident wave incident on a capacitor load, see Fig. 7.21. As in circuits, transient behavior is easily handled in the s-domain. The incident voltage at the input to the line is given by

$$V^+(s) = \frac{2}{s} \frac{Z_o}{R_S + Z_o} = \frac{1}{s}. \tag{7.88}$$

The incident wave reaches the capacitor delayed by d/v_P so that the voltage incident upon the capacitor is given by

$$V_C^+(s) = \frac{e^{-\frac{d}{v_P}s}}{s}. \tag{7.89}$$

The reflection coefficient of the capacitor is given by

$$\Gamma_{\text{CAP}} = \frac{\frac{1}{sC} - Z_o}{\frac{1}{sC} + Z_o} \tag{7.90}$$

so that the reflected wave at the capacitor is given as

$$
\begin{aligned}
V_C^-(s) &= \frac{e^{-\frac{ds}{v_P}}}{s} \frac{\frac{1}{sC} - Z_o}{\frac{1}{sC} + Z_o} = \frac{1 - sCZ_o}{s\left(1 + sCZ_o\right)} e^{-\frac{d}{v_P}s} \\
&= \frac{e^{-\frac{d}{v_P}s}}{s} - \frac{2e^{-\frac{d}{v_P}s}}{s + \frac{1}{Z_oC}}.
\end{aligned} \tag{7.91}
$$

The time-domain form of the reflected voltage at the capacitor is

$$v_C^-(t) = \mathbf{L}^{-1}\left\{V_C^-(s)\right\} = u\left(t - \frac{d}{v_P}\right)\left(1 - 2e^{-\frac{\left(t - \frac{d}{v_P}\right)}{Z_oC}}\right). \tag{7.92}$$

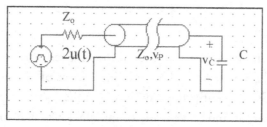

FIGURE 7.21: Unit step wave incident upon a capacitor load.

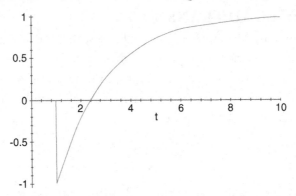

FIGURE 7.22: Reflected voltage at a capacitor load.

The reflected voltage is plotted in Fig. 7.22. At the time of incidence (let's call that $t = 0$), the capacitor is uncharged so that $v_C(0) = 0$ which "looks" like a short circuit. The reflection coefficient of a short circuit is -1 as is the initial portion of the reflected voltage. After a suitable interval, the capacitor is fully charged by the incident signal so that current no longer flows. At this time, the capacitor acts as an open circuit with a reflection coefficient of $+1$ with a corresponding reflected voltage. In spite of the definition of reflection coefficient in terms of frequency-domain impedances, the ratio of the time-domain reflected and incident voltages is often used to describe the reflection coefficient. This more general view corresponds to the reflected voltage when the incident voltage is a unit step. But there are difficulties in applying this to more general wave shapes.

The total voltage across the capacitor is the sum of the incident and the reflected voltages as

$$v_c(t) = v_C^+(t) + v_C^-(t)$$

$$= u\left(t - \frac{d}{v_P}\right) + u\left(t - \frac{d}{v_P}\right)\left(1 - 2e^{-\left(\frac{t - \frac{d}{v_P}}{Z_o C}\right)}\right)$$

$$= 2u\left(t - \frac{d}{v_P}\right)\left(1 - 2e^{-\left(\frac{t - \frac{d}{v_P}}{Z_o C}\right)}\right), \tag{7.93}$$

the usual form for the charging of a capacitor through a resistor. In this case, the effective resistance is Z_o. Since the source has a Thèvenin impedance of Z_o, signals reflected from the capacitor never return. Consequently, the capacitor "sees" Z_o as the impedance connected to it.

In a similar manner, an inductive load, L, initially looks like an open circuit and later, after several time constants, a short circuit. Moreover, it transitions between these two states with a time constant of L/Z_o. The reflected voltage is given by

$$v_L^-(t) = \mathbf{L}^{-1}\left\{V_L^-(s)\right\}$$

$$= u\left(t - \frac{d}{v_P}\right)\left(-1 + 2e^{-\frac{\left(t - \frac{d}{v_P}\right)Z_o}{L}}\right).$$ (7.94)

Similar results can be obtained for impedances with general frequency dependence.

If the source is not matched to the line, then multiple reflections occur. The inclusion of the multiple reflections for frequency-dependent impedances greatly complicates the simple results of the matched source. We will not cover these problems in this textbook.

Time-domain reflectometry (TDR) is the name given to the measurement of the characteristics of transmission systems—lines, waveguides, optical fibers—by the application of pulses and the observation of the reflections. The time delay of the reflection provides information about the location of changes in impedance; the shape and duration of the reflections provides information about the type of impedance discontinuity. This is a very popular measurement technique for finding faults (undesired impedances) on transmission systems. It is also used to characterize the electrical properties of unknown materials.

Example 7.10-1. A TDR system is attached to a 50 Ω transmission line with $v_P = 2 \times 10^8$ m/s to locate a fault. The unit transmits a 1 volt pulse. The reflected signal is shown in Fig. 7.23. Determine the type of fault and its distance from the input of the line. From the return pulse, the down and back round trip time is $2t_d = 12.4$ μs; therefore, the distance of the fault is $d = v_P t_d = 2 \times 10^8(6.2 \times 10^{-6}) = 1240$ m. The ratio of the reflected signal to the transmitted signal is $0.6 = \Gamma_L$. Therefore, $Z_L = Z_o(1 + \Gamma_L)/(1 - \Gamma_L) = 200\ \Omega$.

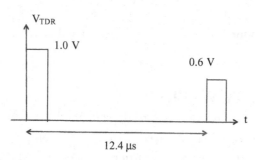

FIGURE 7.23: TDR signals.

Example 7.10-2. Calculate the reflected wave from a parallel RC circuit load on a transmission line of length d when a unit step wave is incident. The transmission line has a characteristic impedance Z_o, the resistor has a value of $R = Z_o$, and the capacitor has a value of C. The incident signal in the s-domain is represented as $V^+ = e^{-ds/v_p}/s$. The impedance of the load is given by $Z_o || 1/sC = Z_o/(1 + sZ_oC)$ so that the reflection coefficient is

$$\Gamma_L = \frac{Z_L - Z_o}{Z_L + Z_o} = \frac{\dfrac{Z_o}{1 + sZ_oC} - Z_o}{\dfrac{Z_o}{1 + sZ_oC} + Z_o} = \frac{-sZ_o^2C}{Z_o(2 + sZ_oC)}$$

$$= \frac{-sZ_oC}{2 + sZ_oC}.$$

The reflected wave is given by

$$V^- = \Gamma_L V^+ = \frac{-sZ_oC}{2 + sZ_oC} \frac{e^{-\frac{ds}{v_P}}}{s} = -\frac{Z_oC e^{-\frac{ds}{v_P}}}{2 + sZ_oC} = \frac{-e^{-\frac{ds}{v_P}}}{s + \dfrac{2}{Z_oC}}$$

which leads to the reflected signal in the time-domain as

$$v^-(t) = -u\left(t - \frac{d}{v_P}\right) e^{-\frac{2\left(t - \frac{d}{v_P}\right)}{Z_oC}}.$$

At the instant of incidence, the reflected wave cancels the incident wave, as it should, since the capacitor appears as a short circuit. But after several time constants, the transient decays to zero since the parallel resistor is matched to the line. The load voltage is given by

$$v_L(t) = v^+(t) + v^-(t) = u\left(t - \frac{d}{v_P}\right)\left(1 - e^{-\frac{2\left(t - \frac{d}{v_P}\right)}{Z_oC}}\right).$$

The load voltage transitions from a short circuit to a matched load.

7.11   STEADY-STATE WAVES

Most communications systems operate in a steady-state mode; the carrier signal is "on" for such a long time that the transients have decayed to zero. For example, at frequencies of 1 GHz during an interval of 1 μs, 10^3 cycles of the carrier signal occur and it is likely that steady-state conditions are established. In contrast to transient waves with many frequency components, steady-state voltages and currents are described by a time-harmonic, single-frequency form, for which it is convenient to use phasor representation. Furthermore, for most applications, we may neglect transmission line losses since they are small, causing only minor differences from lossless line behavior. For lossless conditions, the transmission line propagation constant is given by

$$\gamma = \pm\sqrt{\mathcal{Z}\mathcal{Y}} = \pm\sqrt{(\mathcal{R}+j\omega\mathcal{L})(\mathcal{G}+j\omega\mathcal{C})}$$

$$\xrightarrow[\text{LOSS}\rightarrow 0]{} \pm j\omega\sqrt{\mathcal{L}\mathcal{C}} \qquad (7.95)$$

and the characteristic impedance by

$$Z_C = \sqrt{\frac{\mathcal{Z}}{\mathcal{Y}}} = \sqrt{\frac{\mathcal{R}+j\omega\mathcal{L}}{\mathcal{G}+j\omega\mathcal{C}}} \xrightarrow[\text{LOSS}\rightarrow 0]{} \sqrt{\frac{\mathcal{L}}{\mathcal{C}}}. \qquad (7.96)$$

A marked difference between steady state and transient operation is that steady-state conditions are achieved only after all transients have decayed away. When the sinusoidal generator is first activated, a transient wavefront propagates on the line just as with a pulse generator. With each round trip traversal, the wave undergoes a delay of $2t_d$ and an amplitude multiplication $\Gamma_L\Gamma_S$ due to reflections. After an infinite number of reflections, the transient wave amplitude has decreased to zero and steady-state conditions exist. Careful summation of these infinite waves gives the steady-state wave on the line.

The source located at $z = 0$ excites a wave that propagates to the right, see Fig. 7.24. The phasor form for the wave propagating in the positive z-direction is given by

$$V_0^+(z) = V_o e^{-j\beta z}. \qquad (7.97)$$

After completing one round trip, there is a second wave propagating in the positive z-direction given by

$$V_1^+(z) = V_o e^{-j\beta z}\Gamma_L\Gamma_S e^{-j2\beta d}. \qquad (7.98)$$

After n round trips, the $(n + 1)$th wave in the positive z-direction is given by

$$V_n^+(z) = V_o e^{-j\beta z}\left(\Gamma_L\Gamma_S e^{-j2\beta d}\right)^n. \qquad (7.99)$$

For physically realizable loads and sources, $|\Gamma_L\Gamma_S| < 1$, so that the geometric series of the infinite number of waves can be written in closed form as $\sum a^n = 1/(1-a)$ and the summation

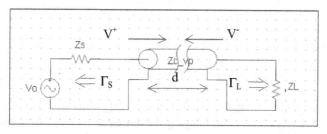

FIGURE 7.24: Sinusoidal excitation of a terminated transmission line.

of these positively propagating voltage waves gives

$$V^+(z) = \sum_{n=0}^{\infty} V_o e^{-j\beta z} \left(\Gamma_L \Gamma_S e^{-j2\beta d}\right)^n$$

$$= V_o e^{-j\beta z} \sum_{n=0}^{\infty} \left(\Gamma_L \Gamma_S e^{-j2\beta d}\right)^n$$

$$= \frac{V_o}{1 - \Gamma_L \Gamma_S e^{-j2\beta d}} e^{-j\beta z}. \qquad (7.100)$$

This is interpreted as a single wave of complex amplitude $V_o/[1 - \Gamma_L \Gamma_S e^{-j2\beta d}]$ propagating toward the load. After establishment of steady-state conditions, an observer "sees" only one wave propagating toward the load! For convenience, define the complex constant associated with this wave as

$$V^+ = \frac{V_o}{1 - \Gamma_L \Gamma_S e^{-j2\beta d}}. \qquad (7.101)$$

In a similar manner, an infinite number of transient waves propagating in the $-z$-direction combine to produce

$$V^-(z) = V_o e^{+j\beta z} \Gamma_L e^{-j2\beta d} \left[1 + \Gamma_L \Gamma_S e^{-j2\beta d} + \left(\Gamma_L \Gamma_S e^{-j2\beta d}\right)^2 + \cdots\right]$$

$$= V_o e^{+j\beta z} \Gamma_L e^{-j2\beta d} \sum_{n=0}^{\infty} \left(\Gamma_L \Gamma_S e^{-j2\beta d}\right)^n$$

$$= V^+ \Gamma_L e^{-j2\beta d} e^{+j\beta z} = V^- e^{+j\beta z}, \qquad (7.102)$$

a single wave of complex amplitude $V^- = V^+ \Gamma_L e^{-j2\beta d}$ propagating toward the source. If $\Gamma_L = 0$, then there is a single wave propagating to the right and no wave to the left. If $\Gamma_S = 0$, then there is a single wave to the right and a single wave to the left. As confirmation, that these two forms are consistent with earlier results, the ratio of the reflected to the incident signal at the load, $z = d$, is

$$\frac{V^-(d)}{V^+(d)} = \frac{V^- e^{+j\beta d}}{V^+ e^{-j\beta d}} = \frac{V^+ \Gamma_L e^{-j2\beta d} e^{+j\beta d}}{V^+ e^{-j\beta d}} = \Gamma_L, \qquad (7.103)$$

the load reflection coefficient. At the input to the line, $z = 0$, the reflected wave is given by

$$V^-(0) = V^+(0) \Gamma_L e^{-j2\beta d}. \qquad (7.104)$$

The incident signal having propagated down the line, reflected from the load, and propagated back to the input of the line. By generalizing the concept of reflection coefficient to be the ratio of the reflected to incident wave at any point on the line, an input reflection coefficient is defined as

$$\Gamma_{IN} = \frac{V^-(0)}{V^+(0)} = \frac{V^-(d)e^{-j\beta d}}{V^+(d)e^{+j\beta d}} = \Gamma_L e^{-j2\beta d}. \qquad (7.105)$$

The incident signal at the input to the line is advanced in phase relative to the incident signal at the load by the electrical length of the line. The reflected signal at the input is retarded in phase relative to the reflected signal at the load by the electrical length of the line. These factors combine to produce a retardation of the reflection coefficient by twice the electrical length of the line. These concepts are the basis for defining the reflection coefficient at a distance d from the load as

$$\Gamma(d) = \Gamma_L e^{-j2\beta d}. \qquad (7.106)$$

The form indicates that as the distance from the load is increased, the reflection coefficient "rotates" in a clockwise (CW) manner in the complex plane. When a distance is reached such that $2\beta d = 2\pi$ the reflection coefficient is the same as at the load. The distance at which this repetition occurs, d_{REPEAT}, is given by

$$d_{REPEAT} = \frac{2\pi}{2\beta} = \frac{\lambda}{2}. \qquad (7.107)$$

The reflection coefficient repeats itself every half-wavelength on the line. When a load Γ_L is attached to a variable length of line, Γ_{IN} can take on any value of the reflection coefficient $|\Gamma_L|e^{j\phi}$ by varying the length of line d over a half-wavelength. This principle is valuable in measurements and in matching techniques that we will cover later.

The phasor nature of the incident and reflected voltages is the key to explaining many other aspects of steady-state transmission line behavior. The phase of the incident wave, $V^+e^{-j\beta z}$, is increasingly negative or retarded with increasing z. An observer moving toward the load would see this as a CW rotation of the incident phasor, see Fig. 7.25. The phase of the reflected wave, $V^-e^{+j\beta z}$, is increasingly positive or advanced with increasing z. An observer moving toward the load would see this as a CCW rotation of the reflected phasor. In Fig. 7.25, the reflected phasor is one-half the incident phasor so that the reflection coefficient has a magnitude of 0.5 everywhere on the line. The uppermost phasor is the incident voltage; the second lower phasor is the reflected voltage; the third lower phasor is the total voltage, i.e., the sum of the incident and reflected voltages; the lowermost phasor is the reflection coefficient. The opposite sense of rotation with distance of the incident and reflected phasors along the transmission line is shown in Fig. 7.25.

The total voltage along the line can be expressed as phasors is key to all behavior which involves both incident and reflected waves. For example, in moving one half-wavelength toward

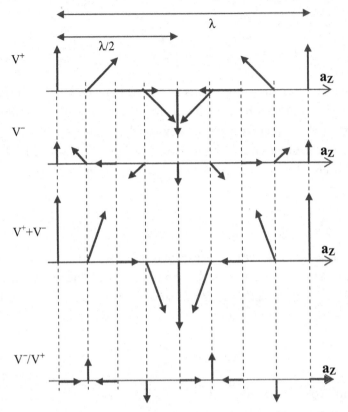

FIGURE 7.25: Variation of incident and reflected phasors with distance along the transmission line.

the load, the incident phasor will change by $-180°$; the reflected phasor will change by $+180°$. The difference between the reflected and the incident as with the reflection coefficient will be $360°$ and will be unaltered.

$$
\begin{aligned}
V(z) &= V^+(z) + V^-(z) \\
&= V^+ \left(e^{-j\beta z} + \Gamma_L e^{-j2\beta d} e^{+j\beta z} \right) \\
&= V^+ e^{-j\beta d} \left(e^{j\beta(d-z)} + \Gamma_L e^{-j\beta(d-z)} \right)
\end{aligned}
\tag{7.108}
$$

where the term $e^{-j\beta d}$ has been factored out so that the distance of the observation point from the load, $d - z$, appears in both the incident and reflected waves. This representation emphasizes the equal and opposite variation with distance of the incident and reflected phasors. Since the phasors rotate oppositely with variations in z, at some points they will be aligned and add and at other points they will be oppositely directed and subtract, see Fig. 7.25. The magnitude of the voltage is expressed as

$$|V(z)| = |V^+(z) + V^-(z)|$$
$$= |V^+ e^{-j\beta z}| |1 + \Gamma_L e^{-j2\beta(d-z)}| \tag{7.109}$$

which is represented more conveniently in normalized form as

$$|V_{\text{NORM}}(z)| = \left| \frac{V(z)}{V^+ e^{-j\beta z}} \right| = |1 + \Gamma_L e^{-j2\beta(d-z)}|$$
$$= |1 + |\Gamma_L| e^{+j\phi_L} e^{-j2\beta(d-z)}|. \tag{7.110}$$

Note that this expression is "fixed" to the transmission line; it is not propagating along the line. The actual voltage on the line varies with time from zero to this peak value, back to zero, then negative to this negative peak value, and back to zero. This is repeated each cycle of the signal.

To help simplify this expression, the complex exponential is converted to real and imaginary parts to obtain the absolute value

$$|V_{\text{NORM}}(z)| = |1 + |\Gamma_L|(\cos[\phi_L - 2\beta(d - z)] + j\sin[\phi_L - 2\beta(d - z)])|$$
$$= \sqrt{1 + |\Gamma_L|^2 + 2|\Gamma_L| \cos[\phi_L - 2\beta(d - z)]}. \tag{7.111}$$

$|V_{\text{NORM}}(z)|$ has maxima of $1 + |\Gamma_L|^2$ at locations where the cosine has a value of $+1$ or

$$(d - z)_{\text{MAX}} = \frac{\phi_L \pm 2n\pi}{2\beta} = \frac{\phi_L \lambda}{4\pi} \pm \frac{n\lambda}{2} \tag{7.112}$$

where $d - z$ is the distance from the maxima to the load. $|V_{\text{NORM}}(z)|$ has minima of $1 - |\Gamma_L|^2$ at locations where the cosine has a value of -1,

$$(d - z)_{\text{MIN}} = \frac{\phi_L \pm (2n - 1)\pi/2}{2\beta} = \frac{\phi_L \lambda}{4\pi} \pm (2n - 1)\frac{\lambda}{4}. \tag{7.113}$$

Since z is less than d, only positive values of n are physically realizable. Plots of $|V_{\text{NORM}}(z)|$ for several different loads are shown in Fig. 7.26. The location of maxima and minima for these loads are summarized in Table 7.1. From these data, it is apparent that the distance to the first maximum increases with increasing phase angle of the reflection coefficient. Note that negative phase angles, $-\phi$, can be expressed as a positive angle, $2\pi - \phi$. The greater the magnitude of the reflection coefficient, the narrower the width of the minima relative to the maxima.

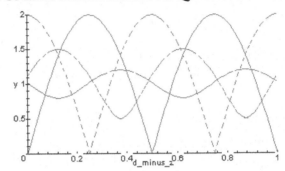

FIGURE 7.26: Normalized voltage magnitude plots for various loads.

The ratio of the voltage maxima to minima is noted as the *voltage standing wave ratio*. *VSWR*, and is expressed as

$$\text{VSWR} = \frac{V_{\text{MAX}}}{V_{\text{MIN}}} = \frac{1 + |\Gamma_L|}{1 - |\Gamma_L|}. \tag{7.114}$$

Voltage maxima exist at locations where the incident and reflected waves are in phase and add; voltage minima are located where they are out of phase and subtract. As with the reflection coefficient, the voltage maxima and minima repeat every $\lambda/2$. Moreover, they are separated from each other by $\lambda/4$. The exact location of maxima or minima depends upon the reflection coefficient as described above.

The greater $|\Gamma_L|$, the greater the VSWR. A small VSWR indicates a small signal reflected. A matched load produces no reflected signal and has a VSWR of 1. When there are no variations in voltage with distance, then the line is called *flat*. VSWR data for Fig. 7.26 are included in Table 7.1.

The accompanying current waves are given by

$$I^+(z) = \frac{V_o^+ e^{-j\beta z}}{Z_o} \tag{7.115}$$

TABLE 7.1: Summary of Standing Wave Plots of Fig. 7.26.

Z_L	Open	Short	$(0.6 + j0.8)Z_C$	$(0.92 - j0.39)Z_C$
Γ_L	$+1$	-1	$0.5\angle\pi/2$	$0.2\angle-\pi/2$
VSWR	∞	∞	3	1.5
$(d-z)_{\text{MAX}}$	$0 + n\lambda/2$	$\lambda/4 + n\lambda/2$	$\lambda/8 + n\lambda/2$	$3\lambda/8 + n\lambda/2$
$(d-z)_{\text{MIN}}$	$\lambda/4 + n\lambda/2$	$0 + n\lambda/2$	$3\lambda/8 + n\lambda/2$	$\lambda/8 + n\lambda/2$

and

$$I^-(z) = -\frac{V_o^+ e^{+j\beta z}}{Z_o}\Gamma_L e^{-j2\beta d}. \tag{7.116}$$

The total current is the sum of the incident and reflected currents as

$$
\begin{aligned}
I(z) &= I^+(z) + I^-(z) \\
&= \frac{V_o^+ e^{+j\beta d}}{Z_o}\left(e^{+j\beta(d-z)} - \Gamma_L e^{-j\beta(d-z)}\right)
\end{aligned}
\tag{7.117}
$$

which compares with Eq. (7.108) except that the reflected wave has a negative sign (recall that the current reflection coefficient is the negative voltage reflection coefficient). This means that the voltage maxima correspond to current minima and vice versa.

At each point along the line, the ratio of the total voltage to the total current can be expressed as an impedance. Due to the presence of the reflected wave, the impedance is not merely the characteristic impedance as it was for transient waves. However, it is easily calculated as the ratio of the total voltage to the total current. The total voltage is composed of the incident and reflected voltage waves; the total current is composed of the incident and reflected current waves. This leads to

$$
\begin{aligned}
Z(z) &= \frac{V(z)}{I(z)} = \frac{V^+(z) + V^-(z)}{I^+(z) + I^-(z)} \\
&= \frac{V^+\left(e^{-j\beta z} + \Gamma_L e^{-j\beta(2d-z)}\right)}{V + \frac{\left(e^{-j\beta z} - \Gamma_L e^{-j\beta(2d-z)}\right)}{Z_o}} \\
&= Z_o\frac{\left(e^{+j\beta(d-z)} + \Gamma_L e^{-j\beta(d-z)}\right)}{\left(e^{+j\beta(d-z)} - \Gamma_L e^{-j\beta(d-z)}\right)}.
\end{aligned}
\tag{7.118}
$$

The impedance depends upon the characteristic impedance, Z_o, the load impedance, Z_L, and the *electrical length* of the line from the observer to the load, $\beta(d - z)$. An alternative viewpoint is to consider the input impedance of a line of length d which is given as

$$
\begin{aligned}
Z_{IN}(d) &= \frac{V_{IN}}{I_{IN}} = \frac{V^+(z=0) + V^-(z=0)}{I^+(z=0) + I^-(z=0)} = Z(z=0) \\
&= Z_o\frac{\left(e^{j\beta d} + \Gamma_L e^{-j\beta d}\right)}{\left(e^{j\beta d} - \Gamma_L e^{-j\beta d}\right)}.
\end{aligned}
\tag{7.119}
$$

In this expression, the argument of Z_{IN} is the length of the line, d, rather than the distance from the input of the line, z. The input impedance varies along the length of the line. In fact, it is a maximum at the location of voltage maxima (and current minima) and minimum at voltage minima (and current maxima). A convenient expression for this variation in terms of load impedance is given as

$$Z_{\text{IN}}(d) = Z_o \frac{\left(e^{j\beta d} + \Gamma_L e^{-j\beta d}\right)}{\left(e^{j\beta d} - \Gamma_L e^{-j\beta d}\right)} = Z_o \frac{\left(e^{j\beta d} + \frac{Z_L - Z_o}{Z_L + Z_o} e^{-j\beta d}\right)}{\left(e^{j\beta d} - \frac{Z_L - Z_o}{Z_L + Z_o} e^{-j\beta d}\right)}$$

$$= Z_o \frac{Z_L \cos\beta d + j Z_o \sin\beta d}{Z_o \cos\beta d + j Z_L \sin\beta d} = Z_o \frac{Z_L + j Z_o \tan\beta d}{Z_o + j Z_L \tan\beta d}. \qquad (7.120)$$

This is a very important result that describes the complex way in which the load impedance is transformed by the line to the input impedance. Let's examine the nature of this transformation more closely.

Of course, a matched load provides an input impedance of Z_o

$$Z_{\text{INMATCH}}(d) = Z_o \frac{Z_o + j Z_o \tan\beta d}{Z_o + j Z_o \tan\beta d} = Z_o. \qquad (7.121)$$

Due to its dependence upon the phasor variations of the incident and reflected waves, the line impedance repeats itself every half wavelength. Accordingly, the input impedance of an integer multiple of a half-wavelength line is the same as the load impedance,

$$Z_{\text{IN}}\left(\frac{n\lambda}{2}\right) = Z_o \frac{Z_L + j Z_o \tan\left(\frac{2\pi}{\lambda} \frac{\lambda}{2}\right)}{Z_o + j Z_L \tan\left(\frac{2\pi}{\lambda} \frac{\lambda}{2}\right)} = Z_L. \qquad (7.122)$$

A quarter-wavelength line provides an inversion of the load impedance as

$$Z_{\text{IN}}\left(\frac{(2n-1)\lambda}{4}\right) = Z_o \frac{Z_L + j Z_o \tan\left(\frac{2\pi}{\lambda} \frac{(2n-1)\lambda}{4}\right)}{Z_o + j Z_L \tan\left(\frac{2\pi}{\lambda} \frac{(2n-1)\lambda}{4}\right)}$$

$$= Z_o \frac{Z_L + j Z_o \tan\left(\frac{\pi}{2}\right)}{Z_o + j Z_L \tan\left(\frac{\pi}{2}\right)} = \frac{Z_o^2}{Z_L}. \qquad (7.123)$$

Short-circuits and open-circuits are relatively easy to make; their input impedances vary as

$$Z_{\text{INSC}}(d) = Z_o \frac{0 + j Z_o \tan\beta d}{Z_o + j0 \tan\beta d} = j Z_o \tan\beta d \qquad (7.124)$$

and

$$Z_{\text{INOC}}(d) = Z_C \frac{\infty + j Z_o \tan\beta d}{Z_o + j\infty \tan\beta d} = \frac{-j Z_o}{\tan\beta d}, \qquad (7.125)$$

respectively. Any reactive value of input impedance can be obtained with a short- or open-circuit on a variable length of line! Plots of $j X_{\text{INSC}}/Z_o$ (solid curve) and $j X_{\text{INOC}}/Z_o$ (dashed

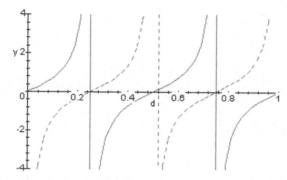

FIGURE 7.27: Input reactance of short-circuit and open-circuit lines.

curve) are shown in Fig. 7.27. As expected, a short-circuit load on a zero length line has zero reactance. For lengths less than $\lambda/4$, the line has a positive reactance and looks inductive; for lengths of $\lambda/4 < d < \lambda/2$, the line looks capacitive; then inductive, then capacitive, ..., changing every $\lambda/4$ as the line length increases. Near the short circuit, the current is large, decreasing with distance from the short, and accompanied by stored magnetic energy. The voltage at the short circuit is zero, increasing with increasing distance from the short circuit. Consequently, there is very little stored electrical energy near the short circuit. The net result is that there is a preponderance of stored magnetic energy near the short circuit as in an inductor. The input impedance has a positive reactance and looks inductive. As the line is lengthened, the voltage increases and the current decreases; the stored magnetic energy decreases and the stored electric energy increases. As the line length approaches $d = \lambda/4$, the total stored magnetic and electric energies become equal. We have seen this before in lumped element circuits; when the stored electric and magnetic energies are equal, the circuit is resonant. Depending upon the configuration, it is either series or parallel resonant. If it is a series resonant circuit, the reactance of the circuit approaches 0; if it is a parallel resonant circuit, the reactance approaches $\pm\infty$. From Fig. 7.27, it is obvious that the input impedance of a short-circuited line looks like a parallel resonant circuit near $\lambda/4$. As the line length increases further for $\lambda/4 < d < \lambda/2$, the large value of voltage and small value of current means that significantly more stored electric energy than magnetic energy is being added. Then, as the length approaches $d = \lambda/2$, the voltage approaches zero and the current approaches a maximum with an increasing preponderance of stored magnetic energy being added. At $d = \lambda/2$, the stored energies are equal so that the reactance becomes zero—the characteristics of a series resonant circuit. Figure 7.28 shows the variation of voltage (solid curve) and current (dashed curve) with distance from the short-circuit load.

From a physical point of view, the results for $d \ll \lambda/4$ is not surprising since a short length of line with a short circuit can be modeled as a one-turn inductor and as such stores

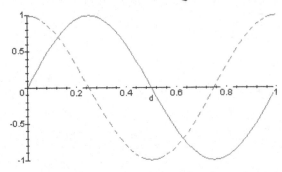

FIGURE 7.28: Current and voltage variations on a line with a short-circuit load.

magnetic energy. But, as the line length increases, the process is more complicated by the presence of both distributed inductive and capacitive effects which cannot be localized. They are distributed throughout the line; their distribution depends upon the load.

The variation of the input reactance of the open-circuit line is very similar. For small lengths of line, it consists of two insulated conductors with a voltage applied across them accompanied by stored electric energy and behaving like a capacitor. But, as the line is lengthened, the voltage at the input to the line becomes smaller and the current becomes larger with an increasing stored magnetic energy added. At $d = \lambda/4$, the stored electric and magnetic energies are equal, the reactance goes to zero, and the input impedance looks like a series resonant circuit. As the line becomes longer still, the inductive effects predominate and the input reactance looks inductive. At $d = \lambda/2$, the stored electric and magnetic energies are equal and the line looks like a parallel resonant circuit. This behavior repeats itself every $\lambda/2$.

The similarity of these effects means that the input reactance of the open- and short-circuited lines are copies of each other. They are merely shifted by $\lambda/4$ to account for the difference between the points of zero voltage and zero current.

This section has focused upon the wave phenomena of transmission lines, but to simplify the analysis and to provide greater insight to the processes, we have modeled the lines with lumped circuit elements. Actually, that is the main reason for invoking the impedance concept with transmission lines since the operation of circuit elements is easier to grasp and to use. Yet, in other cases, the transmission line provides a better or more reliable means of implementing an impedance than a lumped circuit element. This modeling strategy of thinking both lumped element and transmission lines is a standard practice in electromagnetics, but it must be practiced with care. We must not try to stretch the models to encompass applications where the assumptions and limitations that enable their use are invalid. With this word of caution, be prepared for this bold strategy of replacing field and wave concepts by lumped circuit elements wherever possible.

Example 7.11-1. A load of $Z_L = 150 + j50$ Ω is connected to a 2.7 m length of transmission line with $Z_o = 50$ Ω and $v_P = 2 \times 10^8$ m/s. The system operates at a frequency of 100 MHz. Calculate the reflection coefficient of the load, the input reflection coefficient and input impedance, the VSWR, and the location of the voltage maximum and minimum nearest the load. A key to calculation of input parameters is the length of the line in terms of wavelength. This is calculated as $\beta = 2\pi/\lambda = 2\pi f/v_P = \pi$ that leads to $\lambda = v_P/f = 2 \times 10^8/10^8 = 2$ m; the line length is $d = 1.35\lambda$. The load reflection coefficient is calculated as $\Gamma_L = (150 + j50 - 50)/(150 + j50 + 50) = (100 + j50)/(200 + j50) = 0.54e^{+j1.22}$ so that $\Gamma_{IN} = \Gamma_L e^{-j2\beta d} = 0.54e^{+j2.1}$ and $Z_{IN} = Z_o(Z_L + jZ_o\tan\beta d)/(Z_o + jZ_L\tan\beta d) = 31.8e^{+j0.92} = 19.2 + j25.4$ Ω. VSWR $= (1 + |\Gamma_L|)/(1 - |\Gamma_L|) = 3.35$. The voltage maximum nearest the load occurs when $\phi - 2\beta(d - z_{MAX}) = 0$ which leads to $z_{MAX} = d - \phi/2\beta = 2.7 - 0.22/2\pi = 2.66$ m. But this is more than a wavelength from the load. Since voltage maxima occur every half-wavelength, we must subtract a sufficient number of half-wavelengths to keep the result positive, but less than half a wavelength. A subtraction of one wavelength, 2 m, gives $z_{MAX} = 0.66$ m. Since voltage minima occur a quarter-wavelength from maxima, the minimum nearest the load is at $z_{MIN} = 0.16$ m.

Example 7.11-2. Show that for a resistive load that $Z_{INMAX} = Z_o$ VSWR. The input impedance will be maximized at locations of voltage maxima that are also current minima, $Z_{MAX} = V_{MAX}/I_{MIN} = (1 + |\Gamma_L|/[(1 - |\Gamma_L|)/Z_o] = Z_o(1 + |\Gamma_L|/[(1 - |\Gamma_L|) = Z_o$ VSWR.

7.12   STEADY-STATE VOLTAGES AND CURRENTS

The calculation of voltages and currents on transmission lines is relatively easy with the knowledge of transmission line impedance. The inclusion of wave effects in the line impedance enables rather simple KVL and KCL calculations rather than calculations that use wave concepts directly. These ideas are illustrated by calculations for the circuit shown in Fig. 7.29.

The load impedance is expressed as

$$Z_L = R - \frac{1}{j\omega C} = 100 - \frac{j}{2\pi \times 10^8(31.8 \times 10^{-12})}$$

$$= 100 - j50 \text{ Ω} \tag{7.126}$$

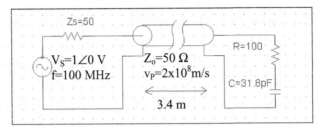

FIGURE 7.29: A transmission line circuit.

which leads to the reflection coefficient as

$$\Gamma_L = \frac{Z_L - Z_o}{Z_L + Z_o} = \frac{100 - j50 - 50}{100 - j50 + 50} = \frac{1-j}{3-j}$$
$$= 0.447e^{-j0.46} = 0.447\angle -26.6°. \tag{7.127}$$

The polar form is usually more convenient than rectangular, especially when using the Smith chart as in the next chapter. Radian measure of the angle is mathematically correct in the exponential form while degrees are commonly used in the magnitude–angle form. The source impedance is matched to the line so that $\Gamma_S = 0$. The wave has a propagation constant of

$$\beta = \frac{\omega}{v_P} = \frac{2\pi \times 10^8}{2 \times 10^8} = \pi \tag{7.128}$$

with a wavelength of

$$\lambda = \frac{2\pi}{\beta} = \frac{2\pi}{\pi} = 2. \tag{7.129}$$

The line length expressed in wavelengths is

$$d_\lambda = \frac{3.4}{2} = 1.7\lambda. \tag{7.130}$$

In the wave point of view, a transient, incident signal is initiated on the line (as with pulses) with an amplitude given by

$$V^+ = 1\frac{Z_o}{Z_o + Z_S} = 0.5 \text{ mV.} \tag{7.131}$$

This wave propagates toward the load, is reflected back to the source, and may be reflected from the source to form a second wave propagating toward the load. In general, this process continues until all the amplitude of the nth wave is negligible; the resulting infinite number of waves can be summed to give a single incident wave as in Eq. (7.101) as

$$V^+(z) = \frac{V_o e^{-j\beta z}}{1 - \Gamma_L \Gamma_S e^{-j2\beta d}} = 0.5 e^{-j\pi z}$$

$$= \frac{V_o e^{-j\beta z}}{1 - \Gamma_L \Gamma_S e^{-j2\beta d}} = 0.5 e^{-j\pi z} \text{ mV}. \tag{7.132}$$

In Fig. 7.29 with a matched source, there is no source reflection coefficient. Therefore, the transient wave is a single incident wave and the calculation is particularly simple. Similarly, the reflected wave can be calculated as

$$V^-(z) = V^+ \Gamma_L e^{-j2\beta d} e^{+j\beta z}$$

$$= 0.5 \left(0.447 e^{-j0.46} \right) e^{-j2\pi(3.4)} e^{+j\pi z}$$

$$= 0.224 e^{-j2.98} e^{+j\pi z} \text{ mV}. \tag{7.133}$$

The input voltage of the line, $z = 0$, is calculated as

$$V_{\text{IN}} = V^+(0) + V^-(0) = 0.5 + 0.224 e^{-j2.98}$$

$$= 0.282 e^{-j0.13} \text{ mV}. \tag{7.134}$$

The input current is given by

$$I_{\text{IN}} = I^+(0) + I^-(0) = \frac{V^+(0) - V^-(0)}{Z_o}$$

$$= \frac{0.5 - 0.224 e^{-j2.98}}{50} = 14.4 e^{j0.053} \text{ }\mu\text{A}. \tag{7.135}$$

The input impedance is given by

$$Z_{\text{IN}} = \frac{V_{\text{IN}}}{I_{\text{IN}}} = \frac{0.282 e^{-j0.136}}{0.0144 e^{+j0.053}} = 19.5 e^{-j0.18}$$

$$= 19.2 - j3.5 \text{ }\Omega. \tag{7.136}$$

Alternatively, the input impedance is calculated directly from Eq. (7.121) as

$$Z_{\text{IN}} = Z_o \frac{Z_L + j Z_o \tan \beta d}{Z_o + j Z_L \tan \beta d}$$

$$= 50 \frac{100 - j50 + j50 \tan(3.4\pi)}{50 + j(100 - j50) \tan(3.4\pi)}$$

$$= 19.5 e^{-j0.18} = 19.2 - j3.5 \text{ }\Omega, \tag{7.137}$$

in agreement with Eq. (7.136). This leads to the equivalent circuit at the input of the transmission line shown in Fig. 7.30.

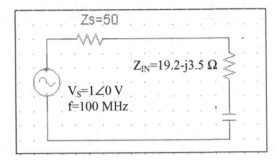

FIGURE 7.30: Input equivalent circuit of the transmission line.

The input voltage and current are calculated from circuit theory as

$$V_{\text{IN}} = V_S \frac{Z_{\text{IN}}}{Z_S + Z_{\text{IN}}} = 0.001 \frac{19.5 e^{-j0.18}}{50 + 19.5 e^{-j0.18}}$$
$$= 0.282 e^{-j0.13} \text{ mV} \qquad (7.138)$$

and

$$I_{\text{IN}} = \frac{V_S}{Z_S + Z_{\text{IN}}} = \frac{0.001}{50 + 19.5 e^{-j0.18}} = 14.4 e^{j0.053} \text{ µA} \qquad (7.139)$$

in agreement with the wave approach.

A third approach is to calculate the input reflection coefficient as

$$\Gamma_{\text{IN}} = \Gamma_L e^{-j2\beta d} = 0.447 e^{-j0.46} e^{-j6.8\pi}$$
$$= 0.447 e^{-j2.98} \qquad (7.140)$$

from which the input impedance is calculated as

$$Z_{\text{IN}} = \frac{V_{\text{IN}}}{I_{\text{IN}}} = Z_o \frac{1 + \Gamma_{\text{IN}}}{1 - \Gamma_{\text{IN}}} = 50 \frac{1 + 0.447 e^{-j2.98}}{1 - 0.447 e^{-j2.98}}$$
$$= 19.5 e^{-j0.18} = 19.2 - j3.5 \ \Omega. \qquad (7.141)$$

The calculation of the load voltage and current are more involved since the incident and reflected voltages transform from the input to the load in with opposite phase variations. The simplest procedure requires calculation of the incident voltage at the source as

$$V^+(0) = \frac{V_{\text{IN}}}{1 + \Gamma_{\text{IN}}} = \frac{0.282^{-j0.13}}{1 + 0.447 e^{-j2.98}} = 0.5 \text{ mV} \qquad (7.142)$$

as in Eq. (7.131). The incident voltage at the load is

$$V^+(d) = V^+(0) e^{-j\beta d} = 0.5 e^{j1.88} \text{ mV} \qquad (7.143)$$

which leads to the load voltage of

$$V_L = V^+(d)[1 + \Gamma_L] = 0.5e^{j1.88}[1 + 0.447e^{-j0.46}]$$
$$= 0.707e^{j1.74} \text{ mV}. \tag{7.144}$$

The load current is calculated as

$$I_L = \frac{V^+(d)[1 - \Gamma_L]}{Z_o} = \frac{0.5e^{j1.88}[1 - 0.447e^{-j0.46}]}{50}$$
$$= 6.3e^{j2.21} \text{ μA}. \tag{7.145}$$

The consistency of the last two calculations is confirmed by the circuit relationship for the load impedance as:

$$Z_L = \frac{V_L}{I_L} = \frac{0.707e^{j1.74}}{0.0063e^{j2.21}} = 100 - j50 \text{ Ω}. \tag{7.146}$$

Since this is a lossless line, no power is lost in the line and the power delivered to the input of the line must equal the power delivered to the load. The power delivered to the input of the line is

$$P_{IN} = Re\left\{\frac{V_{IN}I_{IN}^*}{2}\right\} = Re\left\{\frac{0.282e^{-j0.13}0.01e^{-j0.05}}{2}\right\} = 2 \text{ nW}; \tag{7.147}$$

the power delivered to the load is

$$P_L = Re\left\{\frac{V_L I_L^*}{2}\right\} = Re\left\{\frac{0.707e^{j1.74}0.0063e^{-j2.21}}{2}\right\} = 2 \text{ nW}. \tag{7.148}$$

These procedures are straightforward, though tedious, when executed by hand. Use of complex-math calculators and/or computer applications simplifies them considerably. The Smith chart is a powerful alternative to these methods that offers computational simplification and conceptual insight. That is the focus of the next chapter.

CHAPTER 8

The Smith Chart

8.1   CRANK DIAGRAM

The concept of reflection coefficient is a key detail in the understanding and application of transmission lines. The relationship between the reflected and incident voltage and current waves is readily expressed by the reflection coefficient. However, the concept of impedance is more useful in circuit applications—a more complicated relationship as the ratio of the total voltage and total current. Early transmission line practitioners spent a great deal of effort trying to represent this relationship in a simple, yet illustrative manner. The ratio of the reflected to the incident voltage waves varies with distance from the load as

$$\Gamma(d) = \Gamma_L e^{-j2\beta d}. \tag{8.1}$$

As the distance d from the load to an observer increases, the reflection coefficient is "seen" to rotate in a CW manner. This is due to a more negative phase shift experienced by the reflected wave and less negative phase shift by the incident. This phenomenon is illustrated graphically by assuming an incident wave of unity amplitude and zero degree phase shift, see Fig. 8.1(a). Since the reflected wave is equal to the product of the unity incident wave and the reflection coefficient, the reflected wave is equal to the reflection coefficient

$$V^-(d) = \Gamma(d)V^+(d) = \Gamma_L e^{-j2\beta d}. \tag{8.2}$$

This is represented by the phasor $\Gamma_L e^{-i2\beta d}$ that rotates CW as the distance from the load is increased. The tip of the rotating reflection coefficient phasor traces out the dotted circle of radius $|\Gamma_L|$ in Fig. 8.1(a), completing a revolution every $\lambda/2$. The total voltage is given by

$$V(d) = V^+(d) + V^-(d) = V^+(d)\left[1 + \Gamma_L e^{-j2\beta d}\right]$$
$$= 1 + \Gamma_L e^{-j2\beta d} \tag{8.3}$$

where the incident wave is assumed to be unity with zero phase angle. This is represented by the sum of the incident and reflected phasors in Fig. 8.1(a). Typical phasors are shown for several observation points; the maximum voltage, $V(d_{\text{MAX}})$, occurs where the incident and reflected

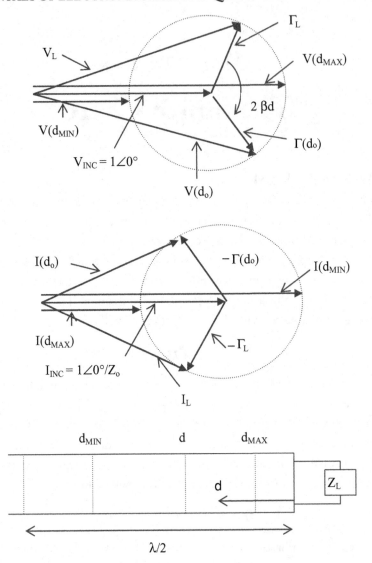

FIGURE 8.1: Incident, reflected, and total phasors on a transmission line: (a) voltage, (b) current, and (c) location.

phasors align; the minimum voltage, $V(d_{MIN})$, occurs where the reflected phasor is oppositely directed to the incident.

For an incident voltage $V^+ = 1$, the incident current is $I^+ = 1/Z_o$ with reflected and total current as

$$I^-(d) = -\Gamma(d)I^+(d) = -\frac{\Gamma_L e^{-j2\beta d}}{Z_o} \tag{8.4}$$

and

$$I(d) = I^+(d) + I^-(d) = I^+(d)\left[1 - \Gamma_L e^{-j2\beta d}\right]$$
$$= \frac{1 - \Gamma_L e^{-j2\beta d}}{Z_o}, \tag{8.5}$$

respectively. The current phasors are displayed in Fig. 8.1(b). Though the incident voltage and current phasors are aligned, the oppositely directed reflected phasors graphically illustrate that voltage maxima and minima are located at the same locations of current minima and maxima, respectively.

The ratio of the voltage to current phasors provides the numeric value of the line impedance as expressed by

$$Z_{\text{IN}}(d) = Z_o \frac{1 + \Gamma_L e^{-j2\beta d}}{1 - \Gamma_L e^{-j2\beta d}} = Z_o \frac{Z_L + jZ_o \tan \beta d}{Z_o + jZ_L \tan \beta d}. \tag{8.6}$$

Variations in the line impedance with distance are due to the rotation of the voltage and current phasors. The impedance values repeat for a given load repeat every $\lambda/2$ as well.

These parameters—the reflection coefficient and the voltage and current phasors—are combined to calculate line impedance in a *Crank diagram* as shown in Fig. 8.2. The reflection coefficient of the load is plotted as a phasor. This phasor is rotated or "cranked" by $e^{-j2\beta d}$ to represent a line of length d. The incident, reflected, and total phasors are sketched for voltage and current. The ratio of the total voltage to current phasors leads to the line impedance a distance d from the load. The variations of this impedance can be calculated for different values of line length. This process provides visualization of transmission line behavior. However, it

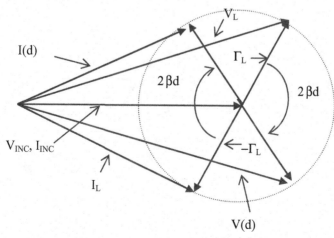

FIGURE 8.2: Crank diagram.

offers little improvement in computational ease. Improvements upon this scheme lead to the Smith chart, the focus of the next section.

Example 8.1-1. Construct the Crank diagram for a load of $Z_L = 100 - j100 \ \Omega$ on a $100 \ \Omega$ line of length $d = 0.2\lambda$. The load reflection coefficient,

$$\Gamma_L = \frac{100 - j100 - 100}{100 - j100 + 100} = \frac{e^{-j1.107}}{\sqrt{5}}$$

$$= 0.447\angle - 63.4°,$$

is plotted in Fig. 8.3. The line length corresponds to a CW rotation of $2\beta d = 2(2\pi/\lambda)(0.2\lambda) = 0.8\pi = 144°$ so that

$$\Gamma(0.2\lambda) = (0.447\angle - 63.4°)(1\angle - 144°)$$

$$= 0.447\angle - 207.4°.$$

With an incident voltage of $1\angle 0°$, $V^+(d) = \Gamma(0.2\lambda)$, and $I^+(d) = -\Gamma(0.2\lambda)/Z_o$, the total voltage is given by $V(0.2\lambda) = 1 + \Gamma(d) = 1 + 0.447\angle - 207.4° = 0.637\angle 18.9°$ and the total current is given by

$$I(0.2\lambda) = \frac{1 - \Gamma(d)}{Z_o} = \frac{1 - 0.447\angle - 207.4}{100} = 0.0146\angle - 8.4°.$$

These phasors are shown in Fig. 8.3. The impedance is calculated as

$$Z(0.2\lambda) = \frac{V(0.2\lambda)}{I(0.2\lambda)} = \frac{0.637\angle 18.9°}{0.0146\angle - 8.4°}$$

$$= 45.1\angle 27.3° = 40.1 + j20.7 \ \Omega.$$

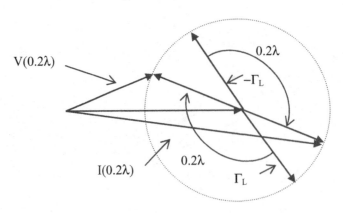

FIGURE 8.3: Crank diagram.

8.2   THE SMITH CHART

The behavior of voltage, current, impedance, and reflection coefficient is graphically illustrated by the Crank diagram. In addition, it offers a systematic, albeit limited, calculation procedure. These ideas form the basis for a much more powerful tool—the Smith chart. In 1939, Philip H. Smith, an engineer at Bell Telephone Laboratories, devised a reflection coefficient chart as a calculation aid for transmission line problems. Though Volpert of the Soviet Union and Mizuhashi of Japan proposed essentially the same chart during the same year, Smith received the recognition.

Quite simply, the Smith chart is a reflection coefficient chart where the complex reflection coefficient is plotted on a set of impedance coordinates. At the heart of the Smith chart is the relationship between an impedance and its reflection coefficient

$$\Gamma = \frac{Z - Z_o}{Z + Z_o} \tag{8.7}$$

and the inverse relationship

$$Z = Z_o \frac{1 + \Gamma}{1 - \Gamma}. \tag{8.8}$$

Smith recognized that all reflection coefficients could be shown on the same chart if the impedances are normalized with respect to the characteristic impedance. The *normalized impedance* is expressed as

$$z = r + jx = \frac{Z}{Z_o} \tag{8.9}$$

so that the reflection coefficient is given as

$$\Gamma = \frac{z - 1}{z + 1} \tag{8.10}$$

and the normalized impedance as

$$z = \frac{1 + \Gamma}{1 - \Gamma}. \tag{8.11}$$

The last two relations are of the general form of the *bilinear transformation*,

$$w = \frac{a + bz}{c + dz}, \tag{8.12}$$

where the complex number z represents impedance or the reflection coefficient and the complex number w represents the other. This defines a unique transformation between the impedance and the reflection coefficient. Actually, the latter transformation is unique only within a range of $0 < d < \lambda/2$ due to the half-wavelength repetition of the reflection coefficient. An interesting

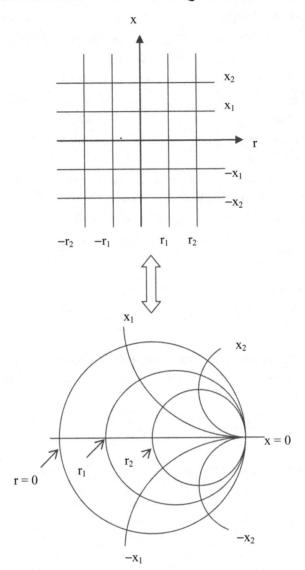

FIGURE 8.4: Transformations between the reflection coefficient and impedance planes.

and useful feature of all bilinear transformations is that circles transform into circles where even a straight line is interpreted as a circle with infinite radius, see Fig. 8.4. Furthermore, the angles between circles in one domain are preserved in the other.

Since $|\Gamma| \leq 1$ for passive loads, all reflection coefficients lie on or within a circle of radius 1. Moreover, every point corresponds to a unique normalized impedance z. The nature of the superimposed impedance coordinates appears complicated, but is really quite simple. It consists

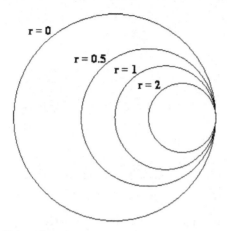

FIGURE 8.5: Reflection coefficient loci for constant resistances.

of two sets of reflection coefficient circles—constant resistance circles and constant reactance circles. To find their equations, the impedance is written in terms of the real and imaginary parts of the reflection coefficient, i.e., $\Gamma = X + jY$, as

$$z = r + jx = \frac{1+\Gamma}{1-\Gamma} = \frac{1+X+jY}{1-X-jY}. \tag{8.13}$$

This is solved for the normalized resistance, r, as

$$r = \frac{1 - X^2 - Y^2}{(1 - X)^2 + Y^2}, \tag{8.14}$$

which is rearranged to

$$\left(X - \frac{r}{1+r}\right)^2 + Y^2 = \left(\frac{1}{1+r}\right)^2. \tag{8.15}$$

This represents circles in the reflection coefficient plane of radius $1/(1+r)$, centered at $X = r/(1+r)$ and $Y = 0$. All circles of constant resistance are symmetrically located with respect to the horizontal axis of the Smith chart. $r = 0$ defines the circle $|\Gamma| = 1$; $r = 1$ passes through the origin which corresponds to $\Gamma = 0$ of a matched load. All values of $r < 1$ lie outside the $r = 1$ circle; all values of $r > 1$ lie inside the $r = 1$ circle. Figure 8.5 shows the loci of Γ in the complex reflection coefficient plane for values of $r = 0, 0.5, 1$, and 2, respectively.

Similarly, the normalized reactance, x, is

$$x = \frac{2Y}{(1 - X)^2 + Y^2}, \tag{8.16}$$

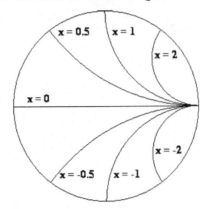

FIGURE 8.6: Reflection coefficient loci for constant reactances.

which is rearranged as

$$(X-1)^2 + \left(Y - \frac{1}{x}\right)^2 = \left(\frac{1}{x}\right)^2. \qquad (8.17)$$

Equation (9.17) represents circles in the reflection coefficient plane of radius $1/x$, centered at $X = 1$ and $Y = 1/x$. All reactance circles are centered on the vertical line tangent to the RHS of the Smith chart; they are tangent to the horizontal midline of the Smith chart. The larger the value of reactance, the smaller the circle; the smaller the value, the larger the circle. The horizontal midline of the chart corresponds to $x = 0$. Positive values of reactance lie in the upper half of the chart; negative values in the lower half. Loci of Γ for values of $x = 0$, ± 0.5, ± 1, and ± 2, respectively, are plotted in Fig. 8.6.

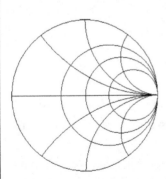

FIGURE 8.7: The Smith chart.

The reflection coefficient loci of Figs. 8.5 and 8.6 are combined for the representation of general impedances, $z = r + jx$. These combined loci provide the impedance-reflection coefficient transformation at the heart of the Smith chart; they are plotted in Fig. 8.7.

The center of the chart represents $\Gamma = 0$, a matched load. The RHS of the horizontal axes represents $\Gamma = +1$, an open circuit. The LHS of the horizontal axis represents $\Gamma = -1$, a short circuit. The outside edge of the chart where $|\Gamma| = 1$ represents purely reactive impedances. Regions outside the chart represent negative resistance, possible only with active devices.

The Smith Chart is now in the public domain. It can be obtained at the following URL, www.utexas.edu. A copy of the Smith Chart from this site is located in Appendix E.

With this foundation, Smith added several useful scales that further enhance the usefulness of the chart. He first added a reflection coefficient magnitude scale. For passive loads, this is just a linear scale for $0 < |\Gamma| < 1$ as the radius of the chart, usually across the bottom of the chart. In fact, the scale is often extended across the entire width of the chart to include the transmission coefficient as $0 < |T| < 2$ since $1 + \Gamma = T$. This scale is most useful when used with a compass. Sometimes there is a closely related scale called *return loss*, usually calibrated in dB, and defined as

$$RL = 10 \log \left| \frac{P_{\text{Refl}}}{P_{\text{Inc}}} \right| = 10 \log |\Gamma|^2 = 20 \log |\Gamma|. \qquad (8.18)$$

Since reflected power is proportional to $|\Gamma|^2$, this is the dB ratio of the reflected to incident power that is "returned" to the source from the load.

Another closely related scale is the VSWR scale utilizing the relationship VSWR $= (1 + |\Gamma|)/(1 - |\Gamma|)$. Since $|\Gamma|$ does not change along a lossless line, the locus of points swept out by the reflection coefficient along the line is a *constant* VSWR *circle*. The VSWR scale, usually along the bottom of the chart, covers the range $1 < \text{VSWR} < \infty$. An interesting feature of a reflection coefficient or VSWR circle is that the value of the normalized resistance where it intersects the positive real axis is equal to the VSWR. No additional scale is needed to compute VSWR.

The real utility of the chart is available with the addition of two, concentric angular scales that encircle the unit outer radius portion of the chart. The first of these is the *reflection coefficient phase angle* measured in degrees. The angle ranges over a full 360° with the positive direction measured CCW from the RH horizontal axis that represents 0°. With this scale and the reflection coefficient magnitude scale, measured or known reflection coefficients can be entered directly on the chart with the corresponding impedance read from the superimposed circles.

The second angular scale represents *electrical length* measured in wavelengths. As the line length is increased, the phase angle of the reflection coefficient becomes increasingly more negative. This angular change in the reflection coefficient is calibrated on the outermost scale in terms of the axial change of position in wavelengths. Since transmission line phenomena repeat every $\lambda/2$, one revolution of the chart corresponds to $\lambda/2$. There are actually two wavelength scales on this portion of the chart. One increases in a CW direction with distance from the load; it actually indicates distance *from* the load, it is denoted as "*toward* generator." The other scale increases in a CCW direction with distance *from* the source; it is denoted as "*toward* load." Both scales begin and end on the LHS of the horizontal scale. In use, the actual numbers on the scale are relatively unimportant; rather the difference between two positions is significant since it corresponds to the line length measured in wavelengths.

In addition to these features, the transform from impedance to reflection coefficient is related to the transform of admittance to the reflection coefficient. The reflection coefficient is expressed as

$$\Gamma = \frac{z-1}{z+1} = \frac{1-\frac{1}{z}}{1+\frac{1}{z}} = \frac{1-y}{1+y} = -\frac{y-1}{y+1}. \tag{8.19}$$

This result indicates that if the normalized admittance is plotted on the Smith chart, then its location corresponds to $-\Gamma$ of the load. This means that the normalized admittance of a load is diametrically opposite the location of the normalized impedance on the Smith chart. This makes calculation of the admittance of the load a simple, graphical procedure—it is the point directly opposite the impedance point on the chart. Moreover, it suggests that the chart can be used interchangeably for impedance or admittance calculations. Finally, the diametrically opposed location of the admittance on the chart can be interpreted as a rotation of the load impedance by $\lambda/4$ which results in an inversion of the impedance. This is a graphical verification of the input impedance of a quarter-wave transformer, see section 9.3.

These details summarize a number of useful features of the Smith chart. But, they have just laid the groundwork for its real utility—finding input impedances for a load on a length of line or load impedances for a given input impedance.

Example 8.2-1. Determine the load impedance and VSWR associated with $\Gamma = 0.5\angle\pi/4$ on a 100 Ω line. The reflection coefficient is plotted on the Smith chart and the normalized impedance is read directly as $z_L = 1.38 + j1.30$, which gives $Z_L = z_L Z_o = 138 + j130\ \Omega$. The constant VSWR circle intersects the positive real axis at $r = 3$, so the VSWR = 3.

Example 8.2-2. A load impedance of $Z_L = 30 - j50\ \Omega$ is attached to a 75 Ω line. Calculate Γ_L, VSWR, and Y_L via the Smith chart. The normalized load impedance $z_L = 0.4 - j0.667$ is plotted on the Smith chart. The reflection coefficient is measured from this point as $\Gamma_L = 0.579\angle - 106°$. The constant VSWR circle intersects the positive real axis at $r = 3.75$ which is also the VSWR. Finally, the z_L is diametrically opposed to z_L and is read off the impedance scales as $y_L = 0.66 + j1.1$, which gives $Y_L = y_L Y_o = y_L/Z_o = 8.8 + j14.7\ \text{mS}$.

Example 8.2-3. Calculate the distance from the load of Example 8.2-2 to the voltage minimum nearest the load. Repeat the calculation for the voltage maximum nearest the load. The angular position of the load on the "toward generator" scale is at 0.397λ. Voltage minima are defined by the intersection of the VSWR circle and the negative real axis of the Smith chart. The distance from the load to the minimum is $d_{\text{MIN}} = 0.103\lambda$. Since voltage maxima are $\lambda/4$ (diametrically opposite) from voltage minima, $d_{\text{MAX}} = 0.353\lambda$.

Example 8.2-4. Calculate the input impedance of a short circuit on a 100 Ω line of 0.2λ length. A short circuit lies on the LHS of the horizontal axis of the Smith chart. The resulting input impedance will also lie on the outer edge of the chart at a distance of 0.2λ CW from the short circuit. The normalized impedance at this point is $z_{IN} = +j3.1$, which corresponds to $Z_{IN} = j310$ Ω corresponding to a lumped element inductor.

8.3   IMPEDANCE TRANSFORMATIONS WITH SMITH CHARTS

The reflection coefficient a distance d from a load is expressed as

$$\Gamma(d) = \Gamma_L e^{-j2\beta d} = \Gamma_L e^{-j\frac{4\pi}{\lambda}d}, \tag{8.20}$$

which corresponds to a CW rotation of the load reflection coefficient. The amount of the rotation is the length of the line d in terms of wavelength. This rotation is accomplished with the aid of the Smith chart to determine the input reflection coefficient and impedance.

The details of this process are best described with a specific example. Consider a load $Z_L = 35 + j75$ Ω connected to a 0.35λ length of 50 Ω line for which the input impedance is needed. This problem can be calculated directly with the Smith chart. The normalized load impedance is $z_L = 0.7 + j1.5$, which is located at $\Gamma_L = 0.67\angle59.9°$ as indicated by the upper-right dot on the Smith chart of Fig. 8.8. The angular position of the load is at 0.167λ on the outer scale "toward generator." The amount of rotation of the reflection coefficient corresponding to the length of the line is 0.35λ. The angular position of the input reflection coefficient is at $0.167 + 0.35 = 0.517λ$. Since the scale extends only to 0.5λ, the actual location corresponds to 0.017λ on the Smith chart. The magnitude of the reflection coefficient doesn't change with rotation, so the radius is located on the same VSWR circle as the load reflection coefficient. The intersection of the VSWR circle and the angular position of the input, denoted by the left dot, is the input reflection coefficient $\Gamma_{IN} = 0.67\angle168°$, which corresponds to a normalized load of $z_L = 0.2 + j0.1$. The input impedance is $Z_{IN} = 10 + j5$ Ω.

This procedure is reversed when the input impedance is known; the rotation is toward the load (CCW) rather than toward the generator (CW). Otherwise, the procedures are identical.

Each of these problems had one impedance and the line length known with the other impedance to be determined. Sometimes, both impedances are known and the line length is to be determined. For valid problems, both impedances must lie on the same VSWR circle. The line length is calculated as the difference between the angular positions of the load impedance

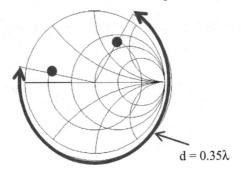

$d = 0.35\lambda$

FIGURE 8.8: Input impedance calculation with the Smith chart.

and the input impedance measured toward the generator (CW). Due to repetition of impedance every $\lambda/2$, this distance is ambiguous by $\pm n\lambda/2$.

The Smith chart makes these calculations this simple! Wow, it's great as a computational aid. In addition, the Smith chart provides a visualization of what occurs on a transmission line. The reflection coefficient phasor rotates in a CW manner with distance from the load toward the generator and vice versa. Moreover, the unique impedance associated with each reflection coefficient is superimposed for easy observation. This is very useful as we design transmission line systems which will optimize power delivered from a source to a load.

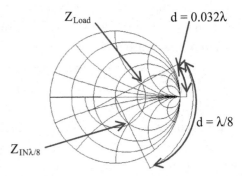

Z_{Load} $d = 0.032\lambda$

$Z_{IN\lambda/8}$

$d = \lambda/8$

FIGURE 8.9: Several calculations with the Smith chart.

Example 8.3-1. Calculate Γ_L and VSWR of $Z_L = 100 + j50\ \Omega$ load on a 50 Ω line. The load is normalized to a value of $z_L = Z_L/Z_o = 2 + j1$. The impedance coordinates for this point are in the upper-right portion of the Smith chart of Fig. 8.9. The reflection coefficient is represented by the phasor from the center of the Smith chart to this point. It has a length of 0.46 (measured on the Γ scale at the bottom of the chart) and an angle of 27° (measured on the innermost angular scale) so that $\Gamma_L = 0.46\angle 27°$. The VSWR is found most simply by rotating the reflection coefficient until its constant VSWR circle intersects the positive, real axis of the chart. The normalized resistance at this point is equal to the VSWR = 2.6. Alternatively, the VSWR scale at the bottom of the Smith chart can be used.

Example 8.3-2. Calculate Y_L for the load of Example 8.3-1. The admittance of an impedance is diametrically opposite as shown in Fig. 8.9. The admittance is $Y_L = y_L Y_o = (0.4 - j0.2)(0.02) = 8 - j4$ mS.

Example 8.3-3. Calculate the distance from the load to the nearest voltage maximum and the distance to the nearest voltage minimum for Example 8.3-1. To reach the voltage maximum, the reflection coefficient must be rotated down to the positive, real axis or a distance of $d_{MAX} = (0.25 - 0.218)\lambda = 0.032\lambda$. The angular distances are read from the outermost scale labeled "toward generator." The distance between maxima and minima is $\lambda/4$ so $d_{MIN} = (0.250 + 0.032)\lambda = 0.282\lambda$.

Example 8.3-4. Calculate the input impedance for the load of Example 8.3-1 at the end of a $\lambda/8$ line. The reflection coefficient must be rotated toward the generator an angular distance of $\lambda/8$ to the location $(0.2180 + 0.125)\lambda = 0.343\lambda$ as indicated in Fig. 8.9. The intersection of this radial line with the constant VSWR circle gives the input impedance as $Z_{IN} = (1 - j1)50 = 50 - j50$ Ω.

8.4   LOSSY TRANSMISSION LINES

To this point, we have considered only lossless transmission lines. However, truely lossless lines do not exist; all physical transmission lines have some loss. However, most useful transmission lines can be considered low-loss. If the lines are not low-loss, they dissipate so much power as signals propagate through the line that they are not useful as transmission lines. By considering only low-loss lines, we observe that the characteristic impedance remains nearly real as described by

$$Z_o \approx \sqrt{\frac{\mathcal{L}}{\mathcal{C}}} \left(1 - \frac{j}{2} \left(\frac{\mathcal{R}}{j\omega\mathcal{L}} + \frac{\mathcal{G}}{j\omega\mathcal{C}} \right) \right) \rightarrow \sqrt{\frac{\mathcal{L}}{\mathcal{C}}}. \tag{8.21}$$

For most transmission lines, the loss is mainly due to copper losses so that $\mathcal{R}$ must be considered, but $\mathcal{G}$ is insignificant. Accordingly, the complex propagation constant is expressed as

$$\gamma = \alpha + j\beta \approx \left(\frac{\mathcal{R}}{2Z_C} + \frac{\mathcal{G}}{2Y_C} \right) + j\omega\sqrt{\mathcal{L}\mathcal{C}}$$

$$\xrightarrow[G=0]{} \frac{\mathcal{R}}{2Z_C} + j\omega\sqrt{\mathcal{L}\mathcal{C}}. \tag{8.22}$$

Substitution of γ for $j\beta$ into Eqs. (8.23) through (8.24) leads to an expression for the incident wave of

$$V^+(z) = \frac{V_o}{1 - \Gamma_L \Gamma_S e^{-2\alpha d} e^{-j2\beta d}} e^{-\alpha z} e^{-j\beta z}$$
$$= V^+ e^{-\alpha z} e^{-j\beta z}. \tag{8.27}$$

The form of this wave is similar to a wave on a lossless transmission line, but the $e^{-\alpha z}$ term describes a wave with amplitude that decreases with distance. The wave is attenuated as it propagates along the transmission line due to the extraction of power from the wave by the losses of the transmission line. In a similar fashion, the reflected wave is given by

$$V^-(z) = V^+ \Gamma_L e^{-2(\alpha+j\beta)d} e^{+\alpha z} e^{+j\beta z} = V^- e^{+\alpha z} e^{+j\beta z} \tag{8.28}$$

where the wave amplitude decreases further as it propagates away from the load. This leads to an input reflection coefficient for a line of length d given by

$$\Gamma_{IN} = \frac{V^-(0)}{V^+(0)} = \frac{V^+ \Gamma_L e^{-2\alpha d} e^{-j2\beta d}}{V^+} = \Gamma_L e^{-2\alpha d} e^{-j2\beta d}. \tag{8.29}$$

The input reflection coefficient is phase shifted from the load as on lossless lines. Additionally, its amplitude is reduced due to the losses of the line. The factor of 2 is due to both down and back reduction in the amplitude as in the phase shift. The effect of a lossy line is to reduce the reflection coefficient; the line appears to be better matched but at the price of power loss. The VSWR varies with position as

$$\text{VSWR}(d) = \frac{1 + |\Gamma(d)|}{1 - |\Gamma(d)|} = \frac{1 + |\Gamma_L| e^{-2\alpha d}}{1 - |\Gamma_L| e^{-2\alpha d}}. \tag{8.30}$$

The input impedance is affected by loss as well according to

$$Z_{IN}(d) = Z_o \frac{Z_L + Z_o \tanh(\gamma d)}{Z_o + Z_L \tanh(\gamma d)}. \tag{8.31}$$

Historically, the attenuation of has been expressed in nepers/m. However, current usage usually expresses the attenuation in terms of dB/m. This expression is formulated as the ratio of the forward power propagating at $z = 1$ m to the forward power propagating at $z = 0$ m as

$$\alpha_{\text{dB/m}} = -10 \log \left(\left| \frac{V^+(d)}{V^+(0)} \right|^2 \right)$$
$$= -10 \log \left(\frac{|V^+(0)|^2 e^{-2\alpha}}{|V^+(0)|^2} \right) = -10 \log \left(e^{-2\alpha} \right)$$
$$= 20\alpha \log (e) = 8.686\alpha. \tag{8.32}$$

A more physical view of the process is obtained by noting that the power transmitted down the line is

$$P(z) = Re\left(\frac{V^+(z)I^+(z)^*}{2}\right) = Re\left(\frac{|V^+|^2 e^{-2\alpha z}}{2 Z_o^*}\right)$$

$$= \frac{|V^+|^2 e^{-2\alpha z}}{2 Z_o} = P(0)e^{-2\alpha z} \tag{8.33}$$

so that

$$\frac{d P(z)}{dz} = \frac{d P(0)e^{-2\alpha z}}{dz} = -2\alpha P(0)e^{-2\alpha z}$$

$$= -2\alpha P(z). \tag{8.34}$$

The attenuation/m is described as

$$\alpha = -\frac{\left(\dfrac{d P(z)}{dz}\right)}{2 P(z)}, \tag{8.35}$$

that is the attenuation is one-half the normalized power lost/meter. This definition provides a means of measuring the attenuation constant α.

Lossy lines can be analyzed on the Smith chart in a manner similar to the lossless lines, but with the reduction of the reflection coefficient with distance from the load. Conversely, the input reflection coefficient grows as the distance to the load is decreased. This is summarized on the left-hand uppermost scale at the bottom of the Smith chart. Each of the bigger marks (yes, it is hard to see a difference between marks unless you look carefully) represents 1 dB. Moving toward the generator reduces the magnitude of the reflection coefficient by the specified dB; moving toward the load increases the magnitude of the reflection coefficient by the specified dB. Alternatively, the magnitude of the lossless reflection coefficient can be directly multiplied by $e^{-2\alpha d}$ or $e^{2\alpha d}$, respectively. This effect is that the locus of the reflection coefficient with distance is no longer a circle, but a spiral—inward for distance toward the generator, outward for distance away from the generator. The simplest procedure is to rotate the reflection coefficient the proper distance as if the line were lossless and then apply the appropriate decrease/increase in the length of the reflection coefficient.

Finally, lossy transmission lines show finite parallel admittance accompanying parallel resonance and nonzero series resistance with series resonance. These values are calculated automatically with Eq. (8.31).

Example 8.4-1. A lossy line with $\gamma = 0.001 + j2\pi$ and $Z_o = 50 \ \Omega$ has a length of 50 m. Calculate the reflection coefficient of a 100 Ω on this line as seen at the load and at the input of the line. The reflection coefficient observed at the load is calculated as $\Gamma_L = (100 - 50)/(100 + 50) = 1/3$. The reflection coefficient observed at the input is calculated as $\Gamma_{IN} = \Gamma_L e^{-2\alpha d} = (1/3)e^{-2(0.001)50} = (1/3)0.904 = 0.302$.

Example 8.4-2. A certain transmission line reduces a signal propagating from the input to the load by a factor of 2. Calculate the reflection coefficient and the VSWR at the line input when a short-circuit load is attached to the line. Assume that the incident signal is 1 V. When it reaches the load it has an amplitude of 0.5 V. After reflection it has an amplitude of −0.5 V. Finally, it propagates back to the input with a further reduction to −0.25 V. The input reflection coefficient is $\Gamma_{IN} = -0.25$ and $\text{VSWR}_{IN} = (1 + 0.25)/(1 - 0.25) = 1.67$. The lossy line reduces the input reflection coefficient and VSWR. A low VSWR on a lossy line indicates that the reflected signal is small, not that a good match has been achieved.

8.5   SLOTTED-LINE MEASUREMENTS

The direct measurement of reflection coefficients with a network analyzer or vector voltmeter has become a standard transmission line technique. It is quick and accurate; with the aid of embedded microprocessors it can make swept frequency measurements with automatic calibration and error correction. But these instruments are expensive! An economical, but more tedious alternative, is the classic *slotted-line* technique with the aid of the Smith chart. The simplicity and elegance of this technique makes it worth our examination.

The heart of this measurement is a section of a slotted transmission line to which the load is attached. The voltage within the slotted line is measured at various axial positions to obtain information about the load. The slotted line is a precision section of the transmission line in which an insulated probe can be inserted to measure the electric fields. Most often the slotted line is composed of a coaxial line mounted in a supporting frame on which a moveable carriage can be slid from end to end. The outer conductor has a narrow, axial slot along its length for access to the fields within the line. The insulated probe, mounted in the carriage, is inserted into the slot in alignment with the electric fields. A voltage is developed between the probe and the outer conductor that is proportional to the electric field and the voltage within the line in the vicinity of the probe. As the carriage is slid along the line, the variations of the voltage within the line can be observed. The design of the slotted line minimizes the effects of the slot on the fields within the line. Moreover, the depth of penetration of the probe into

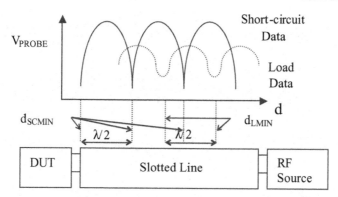

FIGURE 8.10: Slotted line measurement system.

the slot is adjustable to additionally reduce these effects. Typically the probe should be inserted only as far as necessary to obtain acceptable measurements.

Attached to the probe is a diode detector which is intended to operate in the square law region, i.e., the output voltage is proportional to the square of the voltage on the line. Square law operation is typically limited to power levels at the diode of 10 μw or less so the output signal must be greatly amplified. With this in mind, the RF test signal is usually AM modulated by an audio signal. This allows the use of a stable, AC narrowband amplifier tuned to the audio signal instead of a DC amplifier. A diagram of a typical slotted-line measurement system is shown in Fig. 8.10.

The measurement procedure follows. With the load attached to the slotted-line output, the carriage is moved along the line to locate and measure the magnitude and locations of voltage maxima and minima. The locations of voltage minima, d_{LMIN}, are preferred because the voltage minima are narrower than maxima, particularly for high VSWR, and their position can be more accurately determined. Moreover, the probe produces smaller disturbances on the line near minima as explained shortly. The next step is to replace the load with a short circuit and the locations of several short-circuit minima, d_{SCMIN}, are found. Because of the half-wavelength repetition of voltages on lines, d_{SCMIN}, are located an integer number of half-wavelengths from the position of the load. From these measurements, the following information has been obtained $\lambda/2$, VSWR $= V_{\mathrm{MAX}}/V_{\mathrm{MIN}}$, d_{LMIN}, and d_{SCMIN}; the latter is at a point $n\lambda/2$ from the load at which the impedance of load is replicated. The relationship of these data is shown in Fig. 8.10.

The VSWR circle is drawn on the Smith chart as shown in Figure 8.11; the load lies on this circle. A voltage minimum lies at the LH midpoint of this circle corresponding to the location of d_{LMIN}. By rotating from the load voltage minimum to the location of the short-circuit minimum, we arrive at an $n\lambda/2$ location where the line impedance equals the

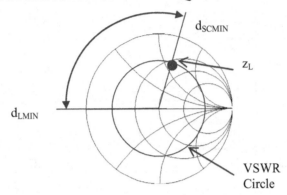

FIGURE 8.11: Slotted-line/Smith chart method for measuring the load impedance.

load impedance. The amount of this rotation is calculated as $|d_{\mathrm{LMIN}} - d_{\mathrm{SCMIN}}|/\lambda$ calculated in wavelengths. The direction of rotation can be either toward the generator or toward the load, whichever is more convenient. See Fig. 8.11 for the Smith chart construction of this procedure.

For uncalibrated sources, the frequency of the generator can be determined by the half-wavelength separation of short-circuit nulls via $f = v_P/\lambda$.

The effects of the probe on the fields within the slotted line can be modeled as an admittance in parallel with the impedance of the line. The real part of the admittance represents the power extracted from the line by the probe and transferred to the measurement system. The imaginary part represents the stored energy in the fields around the probe; usually it is capacitive in nature. An inductor placed in parallel with the probe can create a parallel tuned circuit with the effective capacitance of the probe, producing a very high impedance. This minimizes the "loading" of the line by the probe and it increases the sensitivity of the probe so that it can be even further withdrawn from the slot (thus further reducing its distortion effects on the fields within the slotted line). Unfortunately, the probe capacitance varies with frequency and probe depth requiring a tunable inductor. But, a variable inductor is easy to fabricate at transmission line frequencies; it is just a short circuit on a variable length transmission line. As the length is varied, the inductance varies. This tuning scheme greatly improves the performance of a slotted-line system.

Finally, the use of voltage minima rather than maxima is preferred for another reason. The effective impedance of the probe tends to disturb the fields within the line less near a voltage minimum than near a maximum. The effects of the shunting probe impedance are less when in parallel with the low impedance at a voltage minimum than when in parallel with the high impedance at a voltage maximum.

There are a number of clever, more sophisticated enhancements of this technique, but the fundamentals discussed so far are sufficient for most measurements. But we will continue our focus on the Smith chart in the next chapter as our main tool in the design of matching circuits.

Example 8.5-1. Determine the measurement frequency and the load impedance for the following slotted line measurements: with an unknown load attached a voltage maximum of 4.3 mV is measured; a voltage minimum of 1.6 mV exists at 18.3 cm from the load. With the short circuit attached, the voltage null closest to the load is 25 cm away. The slotted line has a characteristic impedance of 50 Ω and is air-filled. The first null of the short circuit must be a half-wavelength from the load which means that $\lambda = 0.5$ m and $f = v_P/\lambda = 3 \times 10^8/(0.5) = 600$ MHz. The voltage standing wave ratio is VSWR = $4.3/1.6 \approx 2.7$. The rotation from the load voltage minimum to the short-circuit minimum is 6.7 cm = 0.134λ toward the generator. The resulting load impedance is $Z_L = (0.7 + j0.8)50 = 35 + j40$ Ω which compares with the exact solution of $Z_{\text{LEXACT}} = 35.7 + j41.1$ Ω. See Fig. 8.12 for the graphical details of this solution on the Smith chart.

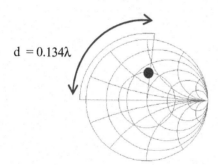

FIGURE 8.12: Smith chart construction for impedance from slotted-line data.

CHAPTER 9

Transmission Line Circuits

9.1   INTRODUCTION

The primary function of transmission lines is to convey signals from one fixed location to another. But while doing so, other characteristics are often exploited. Quite often, the efficiency in the delivery of power to the load is important. Filtering action that passes or blocks certain frequencies is another useful function of transmission lines. In many high-frequency applications, transmission lines operate more nearly as ideal circuit elements than conventional lumped components. This chapter will investigate a several of the more common or important applications of transmission lines. Unfortunately, this is just a sampling of potential uses. You will discover many more in advanced courses or while at work as an engineer. I hope that you will devise a few clever schemes of your own along the way.

9.2   MATCHING CONCEPTS

The efficient transmission of power from a source to a load is crucial in most steady-state transmission line applications. Often the power levels are so low that thermal noise generated within the circuits is as large as the signal. Consequently, it is vital that as much signal power as possible be delivered to the load. With lossless transmission lines, there is no loss within the lines and the net power into one end of a line must exit the other end.

As in circuits, maximum power transfer occurs when $Z_{IN} = Z_S^*$ where Z_{IN} is the input impedance seen by the source "looking" into the transmission line to which the load is attached and Z_S^* is the complex conjugate of the source Thèvenin impedance. The source delivers its available power to the input of the line. All of this power propagates through the lossless line and is delivered to the load. The purpose of all matching schemes is most simply to transform the load impedance Z_L to $Z_{IN} = Z_S^*$ at the transmission line input.

This conjugate match condition holds at every point on a lossless line. This is succinctly stated as, "*on a conjugately matched, lossless transmission line, the impedances seen looking to the right and to the left at every point on the line must be the complex conjugates of each other.*" Mathematically, this means that $Z_{LEFT} = Z_{RIGHT}^*$ and $Y_{LEFT} = Y_{RIGHT}^*$. In addition, for real Z_o, the reflections coefficients are related as $\Gamma_{LEFT} = \Gamma_{RIGHT}^*$, also.

When the source has only a real impedance (as with most commercial generators), i.e., Z_S = real, the line impedance is often matched to the source, $Z_o = Z_S$. In this case, conjugate matching occurs when $Z_{IN} = Z_S^* = Z_S = Z_o$ with $\Gamma_{IN} = 0$ and VSWR = 1. However, when a source with a complex impedance is used, e.g., the output impedance of a transistor, $Z_{IN} = Z_S^* \neq Z_o$, a complex value for which $\Gamma_{IN} \neq 0$ and VSWR > 1. This is not a match as described in earlier chapters since $\Gamma_{IN} \neq 0$. For transients, $\Gamma_{IN} = 0$ is useful since it prevents reflections and preserves the waveshape. But, for steady-state waves with complex loads and sources, $\Gamma_{IN} \neq 0$ is necessary for maximum power transfer.

Different matching designs are based upon achieving this conjugate condition at different points along the line. The choice of different line locations leads to different designs. The choice of design technique depends upon the type of circuit, the environment, and the preferences of the designer. Nevertheless, they all have one common goal—to transform the two terms of impedance, R_L and jX_L, to $R_{IN} = R_S$ and $jX_{IN} = -jX_S$. This procedure requires two variable factors in the matching circuits. Matching circuits are commonly called *tuners*, a term that we will use hereafter. This comes from the fact that the real and reactive parts of the load are "tuned" from original values to proper matching values. The following sections of this chapter will look at a wide variety of matching methods.

9.3 QUARTER-WAVE TRANSFORMER

When the load and source impedances are real, the tuner can be very simple, for example, just a section of line in cascade with the load. The two variable factors in a single transmission line are its characteristic impedance, Z_o, and its length, d. When the line length is chosen as $d = \lambda/4$, the input impedance is given by

$$Z_{IN} = Z_o \left. \frac{Z_L + jZ_o \tan \beta d}{Z_o + jZ_L \tan \beta d} \right|_{d=\frac{\lambda}{4}} = \frac{Z_o^2}{Z_L}. \qquad (9.1)$$

For conjugate match of a real-valued source $Z_{IN} = Z_S^* = Z_S$, the characteristic impedance of the quarter-wave tuner is given by $Z_o = \sqrt{(Z_S Z_L)}$, i.e., the geometric mean of the source and load impedances. With Z_o known, the load impedance can be plotted on the Smith chart. A rotation of this point by 180° CW corresponds to moving toward the source by $\lambda/4$ to the point of the input impedance of the quarter-wave section. The impedance at this point is related to the admittance of the load so that the input impedance of a quarter-wave transformer when normalized with respect to Z_o of the transformer is the same as the normalized load admittance.

This method looks very attractive, but hidden in its simplicity is the difficulty in finding or adjusting the characteristic impedance, Z_o, of coaxial and twisted pair transmission lines. However, It is especially well suited to microstrip and stripline transmission lines where Z_o is related to the width of the ungrounded conductor.

This matching scheme may appear as "magic" since just choosing the proper line can reduce the reflection coefficient and VSWR at the input of the tuner to zero. This means that all of the power incident at the input of the line is delivered to the load. It seems even stranger that this is accomplished by establishing a nonzero VSWR within the quarter-wave section. A closer view of the quarter-wave transformer reveals important details about steady-state matching.

Within the quarter-wave line with a characteristic impedance of $Z_o = \sqrt{(Z_S Z_L)}$ the load presents a reflection coefficient of

$$\Gamma_L = \frac{Z_L - Z_o}{Z_L + Z_o} = \frac{Z_L - \sqrt{Z_L Z_S}}{Z_L + \sqrt{Z_L Z_S}} = \frac{\sqrt{Z_L} - \sqrt{Z_S}}{\sqrt{Z_L} + \sqrt{Z_S}} \qquad (9.2)$$

with a standing wave ratio of

$$\begin{aligned} \text{VSWR} &= \sqrt{\frac{Z_S}{Z_L}}, \qquad \text{for } Z_S > Z_L \\ &= \sqrt{\frac{Z_L}{Z_S}}, \qquad \text{for } Z_S < Z_L. \end{aligned} \qquad (9.3)$$

An incident voltage of $1\angle 0°$ at the input of the quarter-wave line becomes $1\angle 90°$ at the load. When $Z_L > Z_C$, $\Gamma_L = |\Gamma_L| > 0$ so that the reflected voltage at the load is $|\Gamma_L|\angle 90°$ such that $V_{\text{LOAD}} = (1 + |\Gamma_L|\angle 90°)$. The reflected voltage at the input is $|\Gamma_L|\angle 180° = -|\Gamma_L|$ so that the input voltage is $V_{\text{IN}} = 1 - |\Gamma_L| < V_{\text{LOAD}}$. Similarly, the load current is $I_{\text{LOAD}} = (1 - |\Gamma_L|\angle 90°)/Z_o$ and the input current is $I_{\text{IN}} = (1 + |\Gamma_L|)/Z_o > I_{\text{LOAD}}$. Since the input voltage is less than the load voltage and the input current is greater than the load current, the input impedance is less than the load impedance and is given as

$$\begin{aligned} Z_{\text{IN}} &= \frac{V_{\text{IN}}}{I_{\text{IN}}} = Z_o \frac{1 - |\Gamma_L|}{1 + |\Gamma_L|} = \frac{Z_o}{\text{VSWR}} = \frac{\sqrt{Z_S Z_L}}{\sqrt{\frac{Z_L}{Z_S}}} \\ &= Z_S. \end{aligned} \qquad (9.4)$$

In a similar manner, when $Z_S > Z_L$ the input voltage is greater than the load voltage and the input current is less than the load current so that the input impedance is greater than the load impedance according to

$$\begin{aligned} Z_{\text{IN}} &= \frac{V_{\text{IN}}}{I_{\text{IN}}} = Z_C \frac{1 + |\Gamma_L|}{1 - |\Gamma_L|} = Z_C \text{VSWR} \\ &= \sqrt{Z_S Z_L} \sqrt{\frac{Z_S}{Z_L}} = Z_S. \end{aligned} \qquad (9.5)$$

In both cases, the 180° round trip phase shift combined with the proper choice of characteristic impedance produces the desired impedance transformation and a match at the input of the quarter-wave line.

Since $Z_{\text{IN}} = Z_S^* = Z_S$ at the input of the quarter-wave line, there is no reflected wave back toward the source from the quarter-wave line. However, since there is a standing wave within the quarter-wave line, there must be a nonzero reflection coefficient within the line. These two facts seem to be contradictory. In fact, these two facts are consistent, but represent two different viewpoints. Since the load impedance is not equal to the characteristic impedance of the quarter-wave line, there is a reflection coefficient there. As shown in Eq. (9.2), this reflection coefficient is calculated with respect to the characteristic impedance Z_o of the quarter-wave line. The incident and reflected waves within the quarter-wave line combine to produce the input impedance Z_S at the input to the section. This input impedance appears matched to the source and will produce no reflections when it is attached to a line with characteristic impedance Z_S. Calculation of the latter reflection coefficient is made with respect to the characteristic impedance Z_S. Note that the line Z_S used to calculate the reflection coefficient of the input impedance can have any length including zero. These two views give different reflection coefficients because they are calculated with respect to different characteristic impedances, one inside the tuner and one outside.

Though reflection coefficients are a useful way to describe the conditions on a transmission line, they are referenced to a particular impedance level and change as the reference impedance is changed. On the other hand, impedances do not change regardless of the characteristic impedance. For this reason both views are correct—there is no reflected wave at the input of the quarter-wave transformer because standing waves are established within the line which adjust the input impedance to exactly the proper matched value.

The use of reflected waves to alter the impedances on transmission lines is a fundamental principle in matching transmission line circuits. Within a tuner, standing waves exist to alter the impedances, but these effects all combine at the input of the tuner to produce a match. For real sources, this means no reflected power from the tuner back to the source, for complex sources, a conjugate match.

As a matter of interest, quarter-wave transformers are often used in optical equipment. In order to provide an efficient transfer of power from the incident optical wave in air to the lens, a quarter-wave transformer is used. The TEM optical waves behave similarly to transmission line TEM waves. Instead of characteristic impedance, wave impedance is used in calculating the properties of the quarter-wave material according to $\eta_{\lambda/4} = [\eta_{\text{AIR}}\eta_{\text{LENS}}]^{1/2}$. For nonmagnetic materials with $\mu = \mu_o$, this equation becomes $\varepsilon_{\lambda/4} = \varepsilon_0\sqrt{\varepsilon_{\text{RLENS}}}$. To make the coatings work over the range of optical frequencies, multiple quarter-wave transformers are used. The presence of these coatings is observable as the bluish appearance of lenses viewed from an oblique angle.

These coatings enable the optical device to capture all of the incident light, especially important for "night vision" or low light-level applications.

Example 9.3-1. Calculate the characteristic impedance of a quarter-wave transformer used to match a 100 Ω load to a 50 Ω source. From Eq. (8.1), the required characteristic impedance is $Z_o = [(50)(100)]^{1/2} = 70.7$ Ω. This characteristic impedance is not available in coaxial line, but 73 Ω and 75 Ω coax lines can be purchased, probably close enough. Of course the line must be $\lambda/4$ long.

Example 9.3-2. Calculate the required permittivity and thickness of a quarter-wave lens coating for a quartz lens operating at 60 μm wavelength. Since $\varepsilon_{RQUARTZ} = 8.5$ and $\varepsilon_{RAIR} = 1$, the coating material must have $\varepsilon_{RCOATING} = \sqrt{8.5} = 2.92$. The thickness is $d = \lambda/4 = \lambda_o/4/2.92 = 15/2.82 = 5.13$ μm.

9.4 QUARTER-WAVE TRANSFORMERS AND A SINGLE REACTIVE ELEMENT

Quarter-wave transformers provide one method for matching complex loads to real sources as well. The addition of a single reactive element can make the load real with the resultant real component matched to a real source with a quarter-wave transformer. Consider a load $Z_L = 20 - j30$ Ω for discussion of this procedure.

One method is to place an inductive reactance of $jX = 30$ Ω in series with the load to give a combined impedance of $Z_L + jX = 20 - j30 + j30 = 20$ Ω. A quarter-wave transformer of $Z_o = [(20)(50)]^{1/2} = 31.6$ Ω will match this combination to a 50 Ω source.

Alternatively, a parallel inductive reactance $jX = j43.3$ Ω, corresponding to an inductive susceptance of $jB = -j0.0231$ S, cancels the imaginary component of the load admittance, $Y_L = 1/Z_L = 0.0154 + j0.0231$ S. The remaining conductance 0.0154 corresponds to a resistive component of 65 Ω that can be matched to a 50 Ω load with a quarter-wave transformer of $Z_o = [(65)(50)]^{1/2} = 57$ Ω. These two designs are shown in Fig. 9.1. Of course, the inductor values depend upon the frequency of operation as the reactive impedance/admittance due to their frequency dependance.

A somewhat more complicated, but equally valid method, is to connect the load directly to the quarter-wave transformer. The load is transformed to an input impedance of $Z_{IN} = Z_o^2/Z_L = Z_o^2 Y_L = R_{IN} + jX_{IN}$. The real part must be set equal to the source resistance by the choice of Z_o; the imaginary part can be canceled by a series reactive element. In the impedance approach, the load $Z_L = 20 - j30$ Ω is transformed to $Z_{IN} = Z_o^2/(20 - j30) = Z_o^2(0.0154 + j0.0231)$. Since $R_{IN} = 50 = 0.0154Z_o^2$, $Z_o = 57$ ω is required. This results in

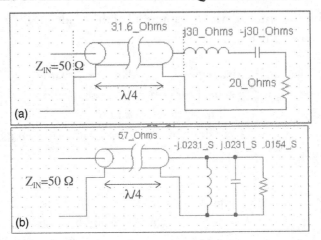

FIGURE 9.1: Two matching schemes for a complex load and a real source.

$jX_{IN} = (57)^2(0.0231) = j75.1$ Ω. The cascade connection of a series capacitive reactance of $-j75.1$ Ω with the input of the quarter-wave line will produce the match.

Alternatively, the input admittance can be expressed as $Y_{IN} = Z_L/Z_o^2 = (20 - j30)/Z_o^2 = G_{IN} + jB_{IN}$. This admittance approach gives $Y_{IN} = (20 - j30)/Z_o^2$ which leads to $20/Z_o^2 = Y_S^* = 0.02$ requiring $Z_o = 31.6$ Ω. A capacitive susceptance of $jB = j30/(31.6)^2 = j0.03$ S must be added in parallel with the input of the quarter-wave transformer to complete the match. These designs are shown in Fig. 9.2.

These first two solutions require inductive elements because the capacitive nature of the load is canceled prior to the impedance transformation by the quarter-wave transformer. A second set of two solutions requires capacitive elements at the load because the load is

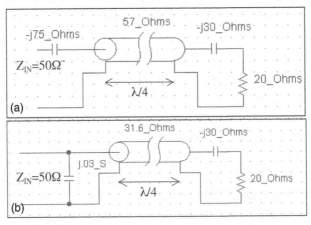

FIGURE 9.2: Two input matching schemes for a complex load and a real source.

transformed before the imaginary term is canceled. Since a quarter-wave transformer moves the load to the opposite side of the Smith chart, this transforms inductive loads to capacitive and vice versa. These examples show that there are a number of options from which to choose the one that best fits the situation.

A different sort of problem results from lumped reactive elements that become less ideal with increased frequency. This is mainly due to increased losses with increased frequency producing a response that is difficult to model. On the other hand, open-circuited and short-circuited transmission lines, hereafter called *stubs*, can provide any desired value of reactance or susceptance with insignificant losses. Consequently, transmission lines are a frequent choice to realize reactive impedances or susceptances at high frequencies.

To illustrate this technique, consider the load $Z_L = 20 - j30$ Ω connected to the 57 Ω quarter-wave transformer in Fig. 9.2(a). A series reactance of $-j75.1$ Ω in cascade with the input of the transformer is required for a match to the 50 Ω source. Open-circuited lines are capacitive for lengths less than $\lambda/4$ with an input impedance of $Z_{INOC} = -j Z_o / \tan \beta d$. Choosing the common value of $Z_o = 50$ Ω, we calculate the required line length from $\tan \beta d = 50/75.1 = 0.666$ as $d = [\tan^{-1}(0.666)]\lambda/2\pi = 0.09\lambda$. The two leads of the line are connected in series with the quarter-wave line as shown in Fig. 9.3(a). Note that this configuration presents a problem when coax line is used as the outer conductor of the stub is ungrounded, an undesired connection. This problem doesn't occur with a twin lead or twisted pair lines where both lines are usually ungrounded.

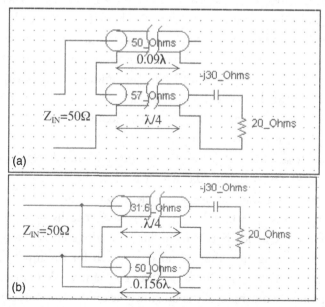

FIGURE 9.3: Matching scheme using transmission line stubs.

A more suitable configuration for a coaxial or microstrip line is the design of Fig. 9.2(b) where a shunt susceptance of $j0.03\ S$ is in cascade with the input of the transformer. This allows one side of the stub to be grounded. Solving $Y_{\text{INSC}} = jY_o \tan \beta d$ for the length of a 50 Ω line, we find $d = [\tan^{-1}(0.03/0.02)]\lambda/2\pi = 0.156\lambda$. This design is shown in Fig. 9.3(b).

In a similar way, inductive elements can be realized with short-circuited lines of length less than $\lambda/4$. The input impedance is $Z_{\text{INSC}} = j Z_o \tan \beta d$ for short-circuited lines.

These examples show just a few of the many possible ways, limited only by the imagination of the designer, to achieve a match. Use the principles described in this section to help you devise some new methods.

A final word of caution—these designs are valid only at a single frequency. The behavior of a particular design at other frequency has not been a concern in this section. Analysis or simulation using SPICE or transmission line analysis software is needed to show the tuner's frequency response.

9.5   SINGLE SUSCEPTANCE TUNERS

The difficulty of fabricating transmission lines with arbitrary characteristic impedance greatly restricts the use of quarter-wave transformers especially for coaxial lines. An alternate, more general strategy is described in this section. This approach is known to transmission line folks as the *single stub tuner* or *single susceptance tuner* (SST). This discussion will be limited to the use of shunt or parallel stubs since this approach will work for coax and microstrip lines.

The basic idea of this approach is to attach the load in cascade with a section of line of proper length to transform the real part of the line's input admittance to the source resistance. An appropriate susceptance is added in parallel to cancel out the line's resulting input susceptance. Though this method is easily accomplished with computers, the Smith chart provides an additional graphical view of the process, hopefully adding insight to the design process. The Smith chart approach will be emphasized here. In describing this procedure, we will match the load of $Z_L = 20 - j30\ \Omega$ to the source with $Z_S^* = 50\ \Omega$.

Using the readily available $Z_o = 50\ \Omega$ for the line and the stub, we normalize the load to $Z_L = (20 - j30)/50 = 0.4 - j0.6$ which is plotted as point A on the Smith chart shown in Fig. 9.4. Since we want to use shunt elements, we must consider the admittance of this load. Its admittance, $y_L = 0.769 + j1.154$, is diametrically opposite the load at point B. The length of line to be added between the load and the stub is determined by the CW rotation, i.e., toward the generator, that produces a match for the input conductance of the line, i.e., $y_{C,D} = 1 \pm jb_L$. The load point moves on a constant VSWR circle with this rotation. The required susceptance at the input of the line must lie on the constant conductance circle of radius 1. The intersection of the constant conductance and constant VSWR circles occurs at the two solutions for the line length. The shorter length of line, $d_{BC} = (0.173 - 0.163)\lambda = 0.01\lambda$, produces a normalized input susceptance of $y_C = 1 + j1.35$ at point C. The longer length of line, $d_{BD} = (0.321 - 0.163)\lambda = 0.158\lambda$, produces an input susceptance of $y_D = 1 - j1.35$ at point D. Note that the two solutions have equal and opposite values of susceptance since the VSWR circle symmetrically intercepts the $g = 1$ circle. This offers a degree of freedom to choose capacitive or inductive susceptance to cancel the remaining susceptance at the input of the line.

The first solution at point C requires a normalized susceptance of $-j1.35$ to cancel the $+j1.35$ input susceptance of the line. This can be realized with a lumped inductor with a value determined by $1/\omega L = 1.35Y_o$ or $L = 1/1.35\omega Y_o = 0.58$ µH for operation on a 50 Ω line at 100 MHz. Note that the normalized admittance was converted back to Siemens, the units of inductive susceptance. A more nearly ideal inductance can be realized with a shorted stub such that $1.35Y_o = 1/Z_o \tan\beta d$ which requires $d = \tan^{-1}(1/1.35)/2\pi = 0.101\lambda$. An alternate method to calculate the stub length is to use the Smith chart directly. The admittance of a short circuit is on the rightmost point of the horizontal axis. As this point is rotated CW toward the source (away from the short circuit), the susceptance becomes decreasingly negative. The required length of the stub to cancel $+j1.35$ susceptance is found at the distance of the point where the input susceptance of the stub is $-j1.35$.

This corresponds to $(0.351 - 0.25)\lambda = 0.101\lambda$. Of course, any $n\lambda/2$ added to the line lies at the same point and is a valid solution. Shorter line lengths are usually more desirable in tuners as we will discuss later. The two inductive designs are shown in Fig. 9.5.

The second possible solution corresponds to the input admittance of the line at point D in Fig. 9.4. This requires a capacitive susceptance of $j1.35$ to cancel the input susceptance of the line. Using similar procedures as in solution for point C, we find that a lumped capacitance defined by $\omega C/Y_o = 1.35$, or $C = 1.35Y_o/\omega = 43$ pF for 100 MHz and 50 Ω. Lumped capacitors are relatively low loss and adjustable models are available for manual "tweaking" by an operator. Alternatively, an open-circuited stub will have a capacitive susceptance and can be used to provide the necessary $+j1.35$ to achieve the match. The required stub length is calculated according to $1.35Y_o = \tan\beta d/Z_o$ or $d = [\tan^{-1}(1.35)]/2\pi = 0.149\lambda$. The admittance of an open circuit lies on the extreme left-hand side of the horizontal axis. This point is rotated a distance of βd to the required admittance $j1.35$;

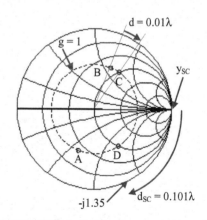

FIGURE 9.4: Smith chart procedures for SST design.

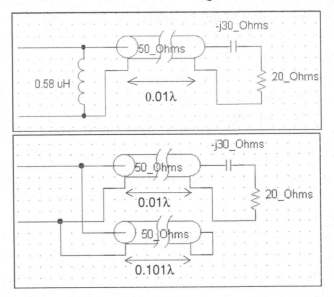

FIGURE 9.5: SST designs with inductive susceptances.

this gives a stub length of $d = (0.149 - 0) = 0.149\lambda$. The two capacitive designs are shown in Fig. 9.6.

In summary, the single susceptance tuner is achieved by attaching the load to a length of line that transforms the load admittance along a constant VSWR circle to the points of input admittance of $1 \pm jb_{IN}$. A parallel susceptance, in lumped element or transmission line form,

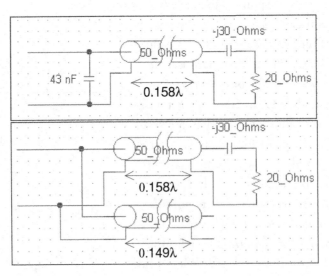

FIGURE 9.6: SST tuners with capacitive susceptances.

with a susceptance of $-[\pm jb_{\rm IN}]$ is added in parallel to the input of this line match the load to the real source.

9.6   GENERAL MATCHING PRINCIPLES

We could delve into a variety of other tuners designs. But, they are based upon two rather general principles. Firstly, the length of a line in cascade with a load can be chosen so that load admittance is transformed (along the constant VSWR circle through the load) so that the real part of the input admittance of the line has the desired value. Secondly, the total susceptance at the input of the line is adjusted by the additional susceptance of a lossless stub in parallel with the input to this line. These two principles will guide you through all tuner design procedures. We will not cover them further in this textbook.

Finally, a few words about the performance of tuners over a range of frequencies. All of the designs above are for a single frequency. The definition of bandwidth for a transmission line circuit is usually stated in terms of the maximum allowed $|\Gamma_{\rm IN}|$ or VSWR. There is no general equation that predicts the performance of all tuners over a range of frequencies. The $|\Gamma_{\rm IN}|$ or VSWR of each tuner must be calculated and plotted over a frequency range to determine the tuner bandwidth. In general, the shorter the line lengths in the matching circuit, the greater the bandwidth. This is due to the variation with frequency of phase differences of the incident and reflected waves. A perfect match is achieved at the center frequency. For all other frequencies, there is a phase difference which is related to $\exp[-j2\Delta\beta d] = \exp[-j2\Delta\omega d/v_P]$. This expression shows that larger d gives greater phase difference for a given change in frequency from the design frequency. The larger the d, the smaller the frequency range over which it is likely that an acceptable match can be achieved.

9.7   CONJUGATE MATCHING

The tuners discussed so far have matched a complex load to a real source that equals the line impedance Z_o. More typically, the source is complex with a real part that is unequal to Z_o. This situation typically occurs when the source is a transistor output. Fortunately, this is not a big problem since the concepts used earlier still work in this case. Firstly, the input conductance is set equal to the desired value, $G_{\rm IN} = G_S \neq Y_C$. Secondly, the input susceptance is not set to zero, but to some finite value, opposite in sign to the source susceptance, i.e., $jB_{\rm IN} = -jB_S$. The general principles remain unchanged—the insertion of a line in cascade with a load changes both the real and imaginary parts of the load admittance; a stub in parallel will alter the susceptance only. The desired input admittance is no longer at the center of the Smith chart, i.e., $\Gamma_{\rm IN} = 0$, rather at some complex value where $Y_{\rm IN} = Y_S^*$.

As an example, consider matching $Z_S = 1 + j1$ to $Z_L = 2 - j2$. To obtain a conjugate match, either we can add something to the source to make it look like at $Z_L^* = 2 + j2$ or we can add something to the load to make it look like $Z_S^* = 1 - j1$. Let's try adding something to the load with an SST type approach to make it look like $Z_S^* = 1 - j1$. As before, we convert to admittances, $y_S = 0.5 - j0.5$ and $y_L = 0.25 + j0.25$, so that we can use parallel elements.

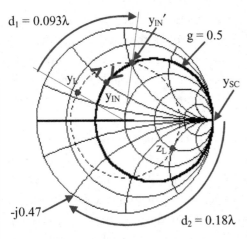

First, add a length of line d_1 by moving along the dashed constant VSWR circle in the direction of the arrow to transform g_L from 0.25 to the required $0.5 = g_S$. From the Smith chart of Fig. 8.7, the line length between the two radials is found to be $d_1 = 0.093\lambda$ and produces $y_{IN}' = 0.5 + j0.97$. Since the input admittance must equal the conjugate of the source admittance, $y_{IN} = y_{STUB} + y_{IN}' = y_S^*$, then $y_{STUB} = 0.5 + j0.5 - 0.5 - j0.97 = -j0.47$. Using a short-circuit stub, we find from the Smith chart that a length of stub $d_2 = 0.18\lambda$ gives the required susceptance in parallel with the cascade line to produce the required conjugate matching. The resulting tuner is shown in Fig. 9.8. As with previ-

FIGURE 9.7: Conjugate matching method.

ous tuner designs, there is another possible solution using the other intercept of the $g = 0.5$ conductance stub. The design of this tuner is left to you.

An alternate design might be to put an SST type of tuner on the generator such that it looks like y_L^*. To do this we need to move CW along a constant VSWR circle from $y_S = 0.5 - j0.5$ via a line length d_1 to the $g = 0.25$ conductance circle. However, this is impossible since the $g = 0.5$ conductance circle upon which y_S lies is located entirely within the $g = 0.25$ conductance circle. The VSWR circle will never intersect the $g = 0.25$ conductance circle. The largest value of conductance which can be achieved by a cascade section of line is

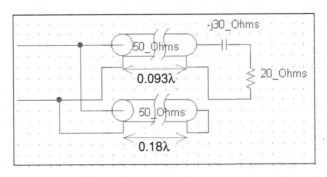

FIGURE 9.8: Conjugately matched tuner.

$g = 0.38$, see Fig. 9.9. This illustrates the fact that there may be "forbidden regions" with some designs to which loads cannot be matched. If this occurs, try another approach.

9.8   ATTENUATORS

The reduction of reflected signals due to lossy elements is exploited by devices called attenuators. As the name implies, attenuators reduce signals, but more importantly, they make a bad mismatch "look" much more like a matched load. This is especially important for some devices which require a low VSWR, but which do not suffer due to a reduction in signal strength.

There are two standard forms for attenuators, pi and tee, as shown in Fig. 9.10. Attenuators are usually specified in terms of the signal reduction that they cause when they are inserted in cascade with a matched load. This is often called the *insertion loss*. By proper choice of components

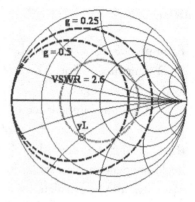

FIGURE 9.9: Scale as appropriate.

an attenuator in a circuit will cause only a negligible reflection when terminated by Z_o. The procedure for designing an attenuator is to consider the series resistance of the attenuator as a resistance/length and the shunt conductance as conductance/length. The square root of their quotient must equal Z_o in order to match the characteristic impedance of the line,

$$Z_o = \sqrt{\frac{R_{\text{SERIES}}}{G_{\text{SHUNT}}}}. \tag{9.6}$$

The amount of attenuation is the other equation that governs the design of attenuators. Attenuators come in a set of rather standard values, 3 dB, 10 dB, and 20 dB.

The main function of attenuators is to reduce the reflection seen by sensitive components by "padding" them with an attenuator, alternatively called a *pad*. Of course this reduces the signal strength of the incident signal by the attenuator rating as well. A short circuit or an open circuit will cause the greatest mismatch, but the lossy nature of the attenuator reduces

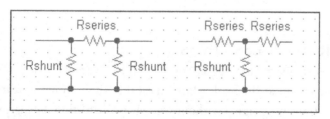

FIGURE 9.10: Pi and tee attenuators.

this significantly. For example, consider that any load on the outer edge of the Smith chart has $|\Gamma| = 1$ and VSWR $= \infty$. With a cascade insertion of 10 dB attenuator this is reduced to a VSWR $= 1.2$. This can be calculated by usual circuit calculations or by using the attenuation scale on the Smith chart. Loads of lesser VSWR are brought even closer to a match by attenuators.

Example 9.8-1. Calculate the resistance values for a 10 dB pi attenuator to be used with a 50 Ω transmission line. The attenuation is calculated with the line impedance attached as a load as shown below.

The input impedance of the attenuator with 50 Ω attached should be 50 Ω. This provides one design equation of $50 = R_{SHUNT} \| (R_{SERIES} + R_{SHUNT} \| 50)$. The power to the load must be reduced by 10 dB from the power into the circuit. Since the input and load impedance are the same, this condition requires that $V_{OUT} = 10^{-10/20} V_{IN} = V_{IN}/\sqrt{10}$. V_{OUT} can also be calculated via a voltage divider equation as

$$V_{OUT} = \frac{R_{SHUNT} \| 50}{R_{SERIES} + R_{SHUNT} \| 50} V_{IN} = \frac{V_{IN}}{\sqrt{10}}.$$

From these two equations, the required resistances are found as $R_{SHUNT} = 96.2$ Ω and $R_{SERIES} = 71.2$ Ω.

Example 9.8-2. Verify the performance of a 3 dB pi attenuator. Using the calculation procedure of Example 9.8-1, we find the values of $R_{SHUNT} = 291.4$ Ω and $R_{SERIES} = 17.7$ Ω. The voltage division of the attenuator is

$$V_{OUT} = \frac{291.4 \| 50}{17.7 + 291.4 \| 50} V_{IN} = 0.707 V_{IN}$$

as it should be. The input impedance is $Z_{IN} = 291.4 \| (17.7 + 291.4 \| 50) = 50$ Ω as expected. The 3 dB pi attenuator works properly!

9.9   DISTORTIONLESS LINES

Preservation of the envelope of many signals propagating on a transmission line is very important for their accurate detection at the load. Mathematical analysis shows that there is a condition for which a line can transmit a signal without waveshape distortion during propagation. Recall that reflections due to mismatches can also cause distortion. Consider that the signal $v(t)$ with a Fourier transform $V(\omega)$ propagates down a line according to

$$V(\omega, z) = V(\omega)e^{-\alpha z}e^{-j\beta z}. \tag{9.7}$$

When α is independent of frequency, this transforms back to the time domain as

$$v(t, z) = v\left(t - \frac{z}{v_P}\right)e^{-\alpha z}, \tag{9.8}$$

the original signal delayed by z/v_P and reduced in amplitude by $\exp[-\alpha z]$ (remember Fourier transforms?). What conditions are necessary to make α independent of frequency? Consider that when (and only when) $\mathcal{R}/\mathcal{L} = \mathcal{C}/\mathcal{G}$, then α becomes

$$
\begin{aligned}
\alpha &= Re\{\sqrt{(R + j\omega L)(G + j\omega C)} \\
&= Re\left\{\sqrt{RG\left(1 + \frac{j\omega L}{R}\right)\left(1 + \frac{j\omega C}{G}\right)}\right\} \\
&= Re\left\{\sqrt{RG}\sqrt{\left(1 + \frac{j\omega L}{R}\right)^2}\right\} = \sqrt{RG}
\end{aligned} \tag{9.9}
$$

which makes α independent of the frequency. Thus, the condition for a distortionless line is that

$$\frac{R}{L} = \frac{G}{C} \tag{9.10}$$

Usually $\mathcal{L}$ is too small for this to be true due to the very small value of $\mathcal{G}$. Though any adjustment of the parameters to bring about the conditions of Eq. (9.10) is possible, it would be foolish to increase $\mathcal{G}$ to make the line distortionless as it would increase α as well. Instead L can be increased, not by altering the line, but by adding lumped inductors in series with the line at intervals that are much less than a wavelength. Since the intervals are so small, the inductance remains essentially distributed and the telegrapher's equations remain true. The line has been made distortionless. In the days when analog long-distance telephony used wire transmission lines across the country, the lines were periodically (with respect to z) "loaded" with series inductance to make the lines distortionless without adding attenuation. Otherwise, the telephone conversations were objectionably distorted.

Example 9.9-1.   A telephone long distance transmission line has $\mathcal{R} = 12.8\ \Omega/km$, $\mathcal{L} = 1.046\ \mu H/km$, $C = 36.4\ \mu F/km$ and $\mathcal{G} = 3.15\ \mu S/km$ for operation up to 3 kHz, the upper limit of the audio band. Calculate the lumped inductance to be added to a line to make it distortionless. From Eq. 9.10, we find that $\mathcal{L}_{REQUIRED} = 12.8(36.4 \times 10^{-9})/3.15 \times 10^{-6} = 91.7\ \mu H/km$ is the required inductance/meter for a distortionless line. For audio operation of this line for which the highest frequency is about 3 kHz, the shortest wavelength is 91 km. With a 450 μH series inductor added every 5 km the wave would see these added inductors as essentially distributed since they are only 0.05λ apart. This gives a total of $1 + 450/5 \sim 90$ μH/km.

9.10 DIRECTIONAL COUPLERS

The slotted line is used to detect the total voltage on a transmission line and from the spatial distribution of the voltage, we can find the VSWR and $|\Gamma_L|$. But it would be more convenient and useful if we could determine the forward and reverse waves directly. A device that will do this is called a *directional coupler*. There are many designs for directional couplers; we will look at a particularly simple one. Consider the circuit shown in Fig. 9.11 where a lumped element resistive network has been inserted in cascade with a line with Z_o.

The voltage across each of the series resistors is expressed in terms of the line current as

$$V_{\text{SERIES}} = R_{\text{SER}} I(z) = \frac{R_{\text{SER}}}{Z_o} \left(V^+ - V^- \right). \tag{9.11}$$

The shunt resistors are chosen so as to make a voltage divider that gives

$$V_{\text{R1}} = \frac{V(z)}{Z_o} = \frac{R_1}{R_1 + R_2} \left(V^+ + V^- \right). \tag{9.12}$$

From these relations, we solve for V_{FOR} as

$$V_{\text{FOR}} = V_{\text{RSERIES}} + V_{\text{R1}}$$
$$= \left(\frac{R_{\text{SER}}}{Z_o} + \frac{R_1}{R_1 + R_2} \right) V^+ + \left(\frac{R_{\text{SER}}}{Z_o} - \frac{R_1}{R_1 + R_2} \right) V^-. \tag{9.13}$$

If we choose $R_{\text{SER}}/Z_o = R_1/(R_1 + R_2)$ then the V^- term vanishes and

$$V_{\text{FOR}} = \left(\frac{R_{\text{SER}}}{Z_o} + \frac{R_1}{R_1 + R_2} \right) V^+ = \frac{2 R_{\text{SER}}}{Z_o} V^+. \tag{9.14}$$

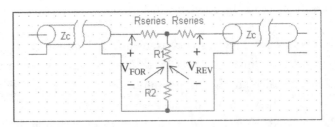

FIGURE 9.11: Directional coupler.

With $R_{SER}/Z_o = R_1/(R_1 + R_2)$, V^- is expressed as

$$V_{REV} = V_{R1} - V_{RSERIES}$$

$$= \left(\frac{R_1}{R_1 + R_2} - \frac{R_{SER}}{Z_o} \right) V^+ + \left(\frac{R_1}{R_1 + R_2} + \frac{R_{SER}}{Z_o} \right) V^-$$

$$= \frac{2 R_{SER}}{Z_o} V^-. \qquad (9.15)$$

The voltages V_{FOR} and V_{REV} can be measured directly and give instantaneous values for the forward and reverse voltages, respectively. VSWR and $|\Gamma_L|$ can be calculated as

$$|\Gamma_L| = \left| \frac{V^-}{V^+} \right| = \left| \frac{V_{REV}}{V_{FOR}} \right| \quad \text{and} \quad \text{VSWR} = \frac{1 + |\Gamma_L|}{1 - |\Gamma_L|}. \qquad (9.16)$$

Best operation is achieved when the directional coupler has the same characteristic impedance as the line, i.e.,

$$\sqrt{2 R_{SER} (R_1 + R_2)} = Z_o, \qquad (9.17)$$

so that it doesn't introduce any reflections due to impedance mismatch. Combining Eq. (9.17) and $R_{SER}/Z_o = R_1/(R_1 + R_2)$, we can solve for R_1 and R_2 under the condition of match to give

$$R_1 = \frac{Z_o}{2} \quad \text{and} \quad R_2 = \frac{Z_o(Z_o - R_{SER})}{2 R_{SER}}. \qquad (9.18)$$

The smaller the value of R_{SER}, the less the attenuation introduced by the directional coupler. But, if it is too small, V_{RSER} may be too small to detect. With $R_{SER} = 1\ \Omega$, a directional coupler matched to a 50 Ω line requires $R_1 = 25\ \Omega$ and $R_2 = 1225\ \Omega$.

Example 9.10-1. Calculate the resistors for a directional coupler to be used on a 300-ohm line. For a reflectionless directional coupler, use Eqs. (9.18). There is no choice in the selection—the first resistor $R_1 = 150\ \Omega$. For convenience, let's use $R_{SER} = 1\ \Omega$ which leads to $R_2 = 300(300 - 1)/2 = 44.85\ \text{k}\Omega$.

Now you have seen a variety of transmission line applications. If you work with transmission lines at high frequencies you will discover many more interesting applications. Maybe a wireless or microwave course is what you need.

Appendix A: Vector Identities

Careful application of associative and distributive properties and the chain rule of differentiation lead to the following vector identities.

$$(\mathbf{A} \times \mathbf{B}) \bullet \mathbf{C} \equiv (\mathbf{B} \times \mathbf{C}) \bullet \mathbf{A} \equiv (\mathbf{C} \times \mathbf{A}) \bullet \mathbf{B}$$

$$\mathbf{A} \times (\mathbf{B} \times \mathbf{C}) \equiv (\mathbf{A} \bullet \mathbf{C}) \mathbf{B} - (\mathbf{A} \bullet \mathbf{B}) \mathbf{C}$$

$$\nabla (\Psi + \Phi) \equiv \nabla \Psi + \nabla \Phi$$

$$\nabla (\Psi \Phi) \equiv \Phi \nabla \Psi + \Psi \nabla \Phi$$

$$\nabla (\mathbf{A} \bullet \mathbf{B}) \equiv (\mathbf{A} \bullet \nabla) \mathbf{B} + (\mathbf{B} \bullet \nabla) \mathbf{A} + \mathbf{A} \times (\nabla \times \mathbf{B}) + \mathbf{B} \times (\nabla \times \mathbf{A})$$

$$\nabla \bullet (\mathbf{A} + \mathbf{B}) \equiv \nabla \bullet \mathbf{A} + \nabla \bullet \mathbf{B}$$

$$\nabla \bullet (\Psi \mathbf{A}) \equiv \mathbf{A} \bullet \nabla \Psi + \Psi \nabla \bullet \mathbf{A}$$

$$\nabla \bullet (\mathbf{A} + \mathbf{B}) \equiv \mathbf{B} \bullet \nabla \times \mathbf{A} - \mathbf{A} \bullet \nabla \times \mathbf{B}$$

$$\nabla \times (\mathbf{A} + \mathbf{B}) \equiv \nabla \times \mathbf{A} + \nabla \times \mathbf{B}$$

$$\nabla \times (\Psi \mathbf{A}) \equiv \nabla \Psi \times \mathbf{A} + \Psi \nabla \times \mathbf{A}$$

$$\nabla \times (\mathbf{A} \times \mathbf{B}) \equiv \mathbf{A} \nabla \bullet \mathbf{B} - \mathbf{B} \nabla \mathbf{A} + (\mathbf{B} \bullet \nabla) \mathbf{A} - (\mathbf{A} \bullet \nabla) \mathbf{B}$$

$$\nabla \bullet \nabla \Psi \equiv \nabla^2 \Psi$$

$$\nabla \times \nabla \times \mathbf{A} \equiv \nabla (\nabla \bullet \mathbf{A}) - \nabla^2 \mathbf{A}$$

$$\nabla \times \nabla \Psi \equiv 0$$

$$\nabla \bullet (\nabla \times \mathbf{A}) \equiv 0$$

Appendix B: Coordinate Systems and Transformations

Three coordinate systems are considered in this text–Cartesian, (circular) cylindrical, and spherical. Transformations of vectors from Cartesian to cylindrical or spherical coordinates and vice-versa are used frequently. Differential lengths, areas, and volumes for these three coordinate systems are required, also. These topics are summarized in this section.

UNIT VECTORS

Each coordinate system requires three mutually orthogonal directions at each point. The direction in which the unit vector points is defined as perpendicular to a surface on which one variable is constant and in the direction of increasing value of the variable. For example, the unit vector $\mathbf{a}_X$ points perpendicularly from an $x = x_o$ plane in the direction of increasing x; the unit vector $\mathbf{a}_\theta$ points perpendicularly from a cone defined by $\theta = \theta_o$ in the direction of increasing θ.

The unit vectors of a coordinate system form a right-handed triad of mutually orthogonal vectors for all three coordinate systems. This means that for unit vectors of the $i, j,$ and k variables that a unit vector has no components in the direction of any other, i.e.,

$$\mathbf{a_i} \bullet \mathbf{a_j} = \mathbf{a_i} \bullet \mathbf{a_k} = \mathbf{a_j} \bullet \mathbf{a_i} = 0.$$

Furthermore, since they are at right angles to each other, the cross product of two is in the direction of the third. When the vectors are properly ordered we have

$$\mathbf{a_i} \times \mathbf{a_j} = \mathbf{a_k}, \quad \mathbf{a_j} \times \mathbf{a_k} = \mathbf{a_i}, \quad \text{and } \mathbf{a_k} \times \mathbf{a_i} = \mathbf{a_j}.$$

The proper order for the vectors is (x, y, z), (ρ, ϕ, z), and (r, θ, ϕ) in Cartesian, cylindrical, and spherical coordinate systems, respectively.

The three coordinate systems are shown in Figure B.1.

COORDINATE TRANSFORMATIONS

Coordinate transformations are accomplished via the dot product; the component of a vector in a particular direction is the dot product of the vector and the unit vector in the particular direction, i.e., $E_i = \mathbf{E} \bullet \mathbf{a}_i$. Each of the components in one coordinate system can be transformed

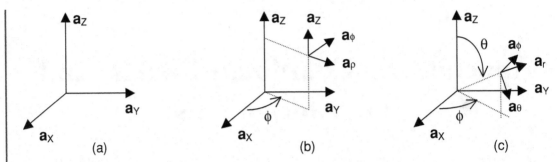

FIGURE B.1: Coordinate Systems; a: Cartesian, b: cylindrical, c: spherical.

to another system by applying this rule. The dot product of the unit vectors can be determined by the geometry of the two vectors.

CARTESIAN (RECTANGULAR) COORDINATES

A vector in Cartesian coordinates is represented as

$$\mathbf{E} = E_X \mathbf{a_X} + E_Y \mathbf{a_Y} + E_Z \mathbf{a_Z} = \sum_{i=1}^{3} E_i \mathbf{a}_i$$

where E_i is the component of $\mathbf{E}$ in the ith direction and $\mathbf{a}_i$ is the unit vector in the ith direction.

(CIRCULAR) CYLINDRICAL COORDINATES

A vector in cylindrical coordinates is represented as

$$\mathbf{E} = E_\rho \mathbf{a}_\rho + E_\varphi \mathbf{a}_\varphi + E_Z \mathbf{a_Z}.$$

The component of $\mathbf{E}$ in a particular direction is formed simply by its dot product with the unit vector in the desired direction as

$$E_\rho = \mathbf{a}_\rho \bullet \mathbf{E} = \mathbf{a}_\rho \bullet (E_X \mathbf{a_X} + E_Y \mathbf{a_Y} + E_Z \mathbf{a_Z}) = E_X \mathbf{a_X} \bullet \mathbf{a}_\rho + E_Y \mathbf{a_Y} \bullet \mathbf{a}_\rho$$
$$= E_X \cos \phi + E_Y \sin \phi,$$
$$E_\varphi = \mathbf{a}_\varphi \bullet \mathbf{E} = \mathbf{a}_\varphi \bullet (E_X \mathbf{a_X} + E_Y \mathbf{a_Y} + E_Z \mathbf{a_Z}) = E_X \mathbf{a_X} \bullet \mathbf{a}_\varphi + E_Y \mathbf{a_Y} \bullet \mathbf{a}_\varphi$$
$$= -E_X \sin \phi + E_Y \cos \phi, \text{ and}$$
$$E_Z = \mathbf{a_Z} \bullet \mathbf{E} = \mathbf{a_Z} \bullet (E_X \mathbf{a_X} + E_Y \mathbf{a_Y} + E_Z \mathbf{a_Z}) = E_Z.$$

The inverse transformations can be accomplished in a similar manner as

$$E_X = \mathbf{a_X} \bullet \mathbf{E} = \mathbf{a_X} \bullet (E_\rho \mathbf{a}_\rho + E_\varphi \mathbf{a}_\varphi + E_Z \mathbf{a_Z}) = E_\rho \mathbf{a}_\rho \bullet \mathbf{a_Z} + E_\varphi \mathbf{a}_\varphi \bullet \mathbf{a_X}$$
$$= E_\rho \cos \phi - E_\varphi \sin \phi,$$

$$E_Y = \mathbf{a}_Y \bullet \mathbf{E} = \mathbf{a}_Y \bullet \left(E_\rho \mathbf{a}_\rho + E_\varphi \mathbf{a}_\varphi + E_Z \mathbf{a}_Z \right) = E_\rho \mathbf{a}_\rho \bullet \mathbf{a}_Y + E_\varphi \mathbf{a}_\varphi \bullet \mathbf{a}_{YX}$$
$$= E_\rho \sin\phi + E_\varphi \cos\phi, \text{ and}$$
$$E_Z = \mathbf{a}_Z \bullet \mathbf{E} = \mathbf{a}_Z \bullet (E_X \mathbf{a}_X + E_Y \mathbf{a}_Y + E_Z \mathbf{a}_Z) = E_Z.$$

SPHERICAL COORDINATES

A vector in spherical coordinates is represented as

$$\mathbf{E} = E_r \mathbf{a}_r + E_\theta \mathbf{a}_\theta + E_\varphi \mathbf{a}_\varphi.$$

The component of $\mathbf{E}$ in a particular direction is formed simply by its dot product with the unit vector in the desired direction as

$$E_r = \mathbf{a}_r \bullet \mathbf{E} = \mathbf{a}_r \bullet (E_X \mathbf{a}_X + E_Y \mathbf{a}_Y + E_Z \mathbf{a}_Z) = E_X \mathbf{a}_X \bullet \mathbf{a}_r + E_Y \mathbf{a}_Y \bullet \mathbf{a}_r + E_Z \mathbf{a}_Z \bullet \mathbf{a}_r$$
$$= E_X \sin\theta \cos\phi + E_Y \sin\theta \sin\phi + E_Z \cos\theta,$$
$$E_\theta = \mathbf{a}_\theta \bullet \mathbf{E} = \mathbf{a}_\theta \bullet (E_X \mathbf{a}_X + E_Y \mathbf{a}_Y + E_Z \mathbf{a}_Z) = E_X \mathbf{a}_X \bullet \mathbf{a}_\theta + E_Y \mathbf{a}_Y \bullet \mathbf{a}_\theta + E_Z \mathbf{a}_Z \bullet \mathbf{a}_\theta$$
$$= E_X \cos\theta \cos\phi + E_Y \cos\theta \sin\phi - E_Z \cos\theta, \text{ and}$$
$$E_\varphi = \mathbf{a}_\varphi \bullet \mathbf{E} = \mathbf{a}_\varphi \bullet (E_X \mathbf{a}_X + E_Y \mathbf{a}_Y + E_Z \mathbf{a}_Z) = E_X \mathbf{a}_X \bullet \mathbf{a}_\varphi + E_Y \mathbf{a}_Y \bullet \mathbf{a}_\varphi + E_Z \mathbf{a}_Z \bullet \mathbf{a}_\varphi$$
$$= -E_X \sin\theta + E_Y \cos\theta.$$

The inverse transformations can be accomplished in a similar manner as

$$E_X = \mathbf{a}_X \bullet \mathbf{E} = \mathbf{a}_X \bullet \left(E_r \mathbf{a}_r + E_\theta \mathbf{a}_\theta + E_\varphi \mathbf{a}_\varphi \right) = E_r \mathbf{a}_r \bullet \mathbf{a}_X + E_\theta \mathbf{a}_\theta \bullet \mathbf{a}_X + E_\varphi \mathbf{a}_\varphi \bullet \mathbf{a}_X$$
$$= E_r \sin\theta \cos\phi + E_\theta \cos\theta \cos\phi - E_\varphi \sin\phi,$$
$$E_Y = \mathbf{a}_Y \bullet \mathbf{E} = \mathbf{a}_Y \bullet \left(E_r \mathbf{a}_r + E_\theta \mathbf{a}_\theta + E_\varphi \mathbf{a}_\varphi \right) = E_r \mathbf{a}_r \bullet \mathbf{a}_Y + E_\theta \mathbf{a}_\theta \bullet \mathbf{a}_Y + E_\varphi \mathbf{a}_\varphi \bullet \mathbf{a}_Y$$
$$= E_r \sin\theta \sin\phi + E_\theta \cos\theta \sin\phi + E_\varphi \cos\phi, \text{ and}$$
$$E_Z = \mathbf{a}_Z \bullet \mathbf{E} = \mathbf{a}_Z \bullet \left(E_r \mathbf{a}_r + E_\theta \mathbf{a}_\theta + E_\varphi \mathbf{a}_\varphi \right) = E_r \mathbf{a}_r \bullet \mathbf{a}_Z + E_\theta \mathbf{a}_\theta \bullet \mathbf{a}_Z + E_\varphi \mathbf{a}_\varphi \bullet \mathbf{a}_Z$$
$$= E_r \cos\theta - E_\theta \sin\theta.$$

DIFFERENTIAL LENGTHS, AREAS, AND VOLUMES

Differential length is defined as a vector which is the sum of the directed differential lengths in each of the three coordinate directions. The equation of the path dictates the unique relationship between the differential lengths. Each directed differential length is defined as the differential length of one of the variables in the direction of the unit vector associated with that variable, i.e., $d\ell_i \mathbf{a}_i$. For coordinates expressed in terms of distance the differential length is obvious. For coordinates expressed in terms of angles, the length requires multiplication of the differential angle by the radius at the point of the differential, i.e., $d\ell_\theta = r d\theta$ or $d\ell_\phi = r \sin\theta d\phi$. The differential lengths form the edges of the differential volume shown in Figure B.2. The directed differential lengths in the three coordinate systems are given as

$$d\ell = dx\mathbf{a}_X + dy\mathbf{a}_Y + dz\mathbf{a}_Z,$$

$$d\ell = d\rho\mathbf{a}_\rho + \rho d\varphi\mathbf{a}_\varphi + dz\mathbf{a}_Z, \text{ and}$$

$$d\ell = dr\mathbf{a}_r + rd\theta\mathbf{a}_\theta + r\sin\theta d\varphi\mathbf{a}_\varphi.$$

The directed differential surface elements, **ds**, are composed of a unit vector perpendicular to the surface multiplied by the differential area of the surface at the point on the surface. The form is given as $\mathbf{ds} = \mathbf{a}_i da_i$ where $\mathbf{a}_i$ is perpendicular to the surface on which the ith variable is constant and da_i is the incremental area on that surface. The direction of the differential element can be defined as dictated by the physical situation in either direction perpendicualar to the surface. These directed surface elements form the six sides of a differential cube as shown in Figure B.2 which can be expressed as

$$\mathbf{ds} = \pm\mathbf{a}_X da_X = \pm\mathbf{a}_X dydz, \ \ \mathbf{ds} = \pm\mathbf{a}_Y da_Y = \pm\mathbf{a}_Y dxdz, \text{ and } \mathbf{ds} = \pm\mathbf{a}_Z da_Z = \pm\mathbf{a}_Z dxdy$$

$$\mathbf{ds} = \pm\mathbf{a}_\rho da_\rho = \pm\mathbf{a}_\rho \rho d\varphi dz, \ \ \mathbf{ds} = \pm\mathbf{a}_\varphi da_\varphi = \pm\mathbf{a}_\phi d\rho dz, \text{ and}$$

$$\mathbf{ds} = \pm\mathbf{a}_Z da_Z = \pm\mathbf{a}_Z \rho d\phi d\rho \text{ and}$$

$$\mathbf{ds} = \pm\mathbf{a}_r da_r = \pm\mathbf{a}_r r\sin\theta d\theta d\varphi, \ \ \mathbf{ds} = \pm\mathbf{a}_\theta da_\theta = \pm\mathbf{a}_\theta r\sin\theta d\varphi dr, \text{ and}$$

$$\mathbf{ds} = \pm\mathbf{a}_\varphi da_\varphi = \pm\mathbf{a}_\varphi r d\theta dr.$$

The differential volumes are formed as the product of the three differential lengths which define the cubes as shown in Figure B.2. The differential volumes are expressed as

$$dv = dxdydz, \ \ dv = \rho d\rho d\varphi dz, \text{ and } dv = r^2 \sin\theta dr d\theta d\varphi.$$

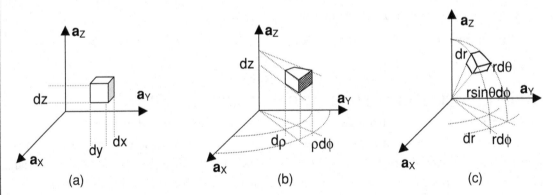

(a)   (b)   (c)

FIGURE B.2: Differential Surface and Volume Elements a: Cartesian, b: cylindrical, c: spherical.

Appendix C: Vector Calculus

DIFFERENTIAL OPERATIONS

The fundamental vector differential operators are given in Cartesian, cylindrical, and spherical coordinate systems.

Gradient

Cartesian

$$\nabla V = \frac{\partial V}{\partial x}\mathbf{a}_X + \frac{\partial V}{\partial y}\mathbf{a}_Y + \frac{\partial V}{\partial z}\mathbf{a}_Z$$

Cylindrical

$$\nabla V = \frac{\partial V}{\partial \rho}\mathbf{a}_\rho + \frac{1}{\rho}\frac{\partial V}{\partial \phi}\mathbf{a}_\varphi + \frac{\partial V}{\partial z}\mathbf{a}_Z$$

Sperical

$$\nabla V = \frac{\partial V}{\partial r}\mathbf{a}_r + \frac{1}{r}\frac{\partial V}{\partial \theta}\mathbf{a}_\theta + \frac{1}{r\sin\theta}\frac{\partial V}{\partial \phi}\mathbf{a}_\varphi$$

Divergence

Cartesian

$$\nabla \bullet \mathbf{D} = \frac{\partial D_X}{\partial x} + \frac{\partial D_Y}{\partial y} + \frac{\partial D_Z}{\partial z}$$

Cylindrical

$$\nabla \bullet \mathbf{D} = \frac{1}{\rho}\frac{\partial (\rho D_\rho)}{\partial \rho} + \frac{1}{\rho}\frac{\partial D_\varphi}{\partial \varphi} + \frac{\partial D_Z}{\partial z}$$

Spherical

$$\nabla \bullet \mathbf{D} = \frac{1}{r^2}\frac{\partial (r^2 D_r)}{\partial r} + \frac{1}{r\sin\theta}\frac{\partial (\sin\theta D_\theta)}{\partial \theta} + \frac{1}{r\sin\theta}\frac{\partial D_\varphi}{\partial \varphi}$$

Curl

Cartesian

$$\nabla \times \mathbf{H} = \left(\frac{\partial H_Z}{\partial y} - \frac{\partial H_Y}{\partial z}\right)\mathbf{a}_X + \left(\frac{\partial H_X}{\partial z} - \frac{\partial H_Z}{\partial x}\right)\mathbf{a}_Y$$

$$+ \left(\frac{\partial H_Y}{\partial x} - \frac{\partial H_X}{\partial y}\right)\mathbf{a}_Z$$

Cylindrical

$$\nabla \times \mathbf{H} = \left(\frac{1}{\rho} \frac{\partial H_Z}{\partial \varphi} - \frac{\partial H_\varphi}{\partial z} \right) \mathbf{a}_\rho + \left(\frac{\partial H_\rho}{\partial z} - \frac{\partial H_Z}{\partial \rho} \right) \mathbf{a}_\varphi$$

$$+ \frac{1}{\rho} \left(\frac{\partial (\rho H_\varphi)}{\partial \rho} - \frac{\partial H_\rho}{\partial \varphi} \right) \mathbf{a}_Z$$

Spherical

$$\nabla \times \mathbf{H} = \frac{1}{r \sin \theta} \left(\frac{\partial (\sin \theta H_\varphi)}{\partial \theta} - \frac{\partial H_\theta}{\partial \varphi} \right) \mathbf{a}_r$$

$$+ \frac{1}{r} \left(\frac{1}{\sin \theta} \frac{\partial H_r}{\partial \varphi} - \frac{\partial (r H_\varphi)}{\partial r} \right) \mathbf{a}_\theta$$

$$+ \frac{1}{r} \left(\frac{\partial (r H_\theta)}{\partial r} - \frac{\partial H_r}{\partial \theta} \right) \mathbf{a}_\varphi$$

Laplacian

Cartesian

$$\nabla^2 V = \frac{\partial^2 V}{\partial x^2} + \frac{\partial^2 V}{\partial y^2} + \frac{\partial^2 V}{\partial z^2}$$

Cylincrical

$$\nabla^2 V = \frac{1}{\rho} \frac{\partial V}{\partial \rho} \left(\rho \frac{\partial V}{\partial \rho} \right) + \frac{1}{\rho^2} \frac{\partial^2 V}{\partial \varphi^2} + \frac{\partial^2 V}{\partial z^2}$$

Spherical

$$\nabla^2 V = \frac{1}{r^2} \frac{\partial}{\partial r} \left(r^2 \frac{\partial V}{\partial r} \right) + \frac{1}{r^2 \sin \theta} \frac{\partial}{\partial \theta} \left(\sin \theta \frac{\partial V}{\partial \theta} \right)$$

$$+ \frac{1}{r^2 \sin^2 \theta} \frac{\partial^2 V}{\partial \varphi^2}$$

INTERGRAL OPERATIONS
Divergence Theorem
The integral of the normal component of any vector field over a closed surface is equal to the integral of the divergence of this vector field throughout the volume enclosed by the closed surface if the derivatives of the divergence are finite. The surface S completely encloses the volume V. **ds** is taken to be the outward normal differential surface element.

$$\oint_S \mathbf{D} \cdot \mathbf{ds} = \iiint_V \nabla \cdot \mathbf{D} dV$$

Stokes Theorem

The line integral of a vector field along a closed path is equal to the integral of the normal of the curl of this vector field over any surface whose perimeter is defined by the path of the line integral. The right hand rule relates the surface normal of the enclosed surface S to the direction of the path L.

$$\oint_L \mathbf{E} \bullet d\ell = \iint_S (\nabla \times \mathbf{E}) \bullet \mathbf{ds}$$

INTEGRAL EVALUATION

Line (Work) Integrals

Line integrals of the form $\int_L \mathbf{E} \bullet d\ell$ are proportional to work done by movement along the directed path L. Their evaluation is easiest when executed in an appropriate coordinate system. For example, if the path is on a circular arc centered on the z-axis and lying in a constant z plane, the cylindrical coordinate system is more appropriate than Cartesian. For more information regarding $\mathbf{d}\ell$ refer to Appendix B. The directed differential length $\mathbf{d}\ell$ can be expressed as

$$\text{Cartesian} \qquad \mathbf{d}\ell = dx\mathbf{a}_X + dy\mathbf{a}_Y + dz\mathbf{a}_Z$$

$$\text{Cylindrical} \qquad \mathbf{d}\ell = d\rho\mathbf{a}_\rho + \rho d\varphi\mathbf{a}_\varphi + dz\mathbf{a}_Z$$

$$\text{Sphere} \qquad \mathbf{d}\ell = dr\mathbf{a}_r + rd\theta\mathbf{a}_\theta + r\sin\theta d\varphi\mathbf{a}_\varphi$$

in the common coordinate systems. When the integration path is along a single coordinate then all other differentials are zero. When the integration path is defined by equations, the differentials are related to each other by the equations of the path. For example, if the path follows the parabolic curve, $y = x^2$, $z = 2$ (remember that it takes two equations to define a path) then $dy = 2xdx$ relates the two differentials while dz is independent. The evaluation of the line integrals is expedited by substituting these relations into the scalar integrals so that the expressions are in terms of a single variable and can be evaluated directly. For example, if $\mathbf{E} \bullet d\ell = ydx$ on a path defined by $y = x^2$ from (0,0) to (2,4), then we could substitute for y to obtain $ydx = x^2 dx$ which can be integrated from $x = 0$ to $x = 2$. Alternatively, we could substitute $dx = dy/2x = dy/2\sqrt{y}$ to obtain $ydx = \sqrt{y}dy/2$ integrated from $y = 0$ to $y = 4$. The results are identical.

Surface (Flux) Integrals

Surface integrals of the form $\iint_S \mathbf{D} \bullet \mathbf{ds}$ describe the total flux which passes through surface S. Their evaluation is easiest when executed in a coordinate system which matches the geometry of the boundary. For example, it is more appropriate to use spherical rather than Cartesian coordinates to calculate the flux through a sphere of radius a; Cartesian coordinates are a better choice than spherical when the flux through a planar surface is calculated. The unit normal to the surface S provides the direction of **ds**; the surface element perpendicular to the normal da provides the magnitude. The differential lengths which make up da lie on the surface S. For more details regarding **ds**, refer to Appendix B. Analytic evaluation of surface integrals which do not conform to one of these coordinate systems are beyond the scope of this text; numeric evaluation can be used for such surfaces. The directed differential surface element **ds** can be expressed as

Cartesian:

$$\mathbf{ds} = \pm \mathbf{a}_X da_X = \pm \mathbf{a}_X dy\,dz, \quad \mathbf{ds} = \pm \mathbf{a}_Y da_Y = \pm \mathbf{a}_Y dx\,dz, \quad \text{and } \mathbf{ds} = \pm \mathbf{a}_Z da_Z = \pm \mathbf{a}_Z dx\,dy$$

Cylindrical:

$$\mathbf{ds} = \pm \mathbf{a}_\rho da_\rho = \pm \mathbf{a}_\rho \rho\,d\varphi\,dz, \quad \mathbf{ds} = \pm \mathbf{a}_\varphi da_\varphi = \pm \mathbf{a}_\phi d\rho\,dz, \quad \text{and } \mathbf{ds} = \pm \mathbf{a}_Z da_Z = \pm \mathbf{a}_Z \rho\,d\phi\,d\rho$$

Spherical:

$$\mathbf{ds} = \pm \mathbf{a}_r da_r = \pm \mathbf{a}_r r \sin\theta\,d\theta\,d\varphi, \quad \mathbf{ds} = \pm \mathbf{a}_\theta da_\theta = \pm \mathbf{a}_\theta r \sin\theta\,d\varphi\,dr, \quad \text{and}$$
$$\mathbf{ds} = \pm \mathbf{a}_\varphi da_\varphi = \pm \mathbf{a}_\varphi r\,d\theta\,dr$$

Appendix D: Material Properties

CONDUCTIVITY

The conductivities for a variety of common conductive materials are given in Table D.1. These values are typical at room temperatures and for low frequencies. More detailed specifications can be found in a variety of handbooks.

TABLE D.1

MATERIAL (METALS)	$\sigma - (\Omega M)^{-1}$
Perfect Electric Conductor (PEC)	∞
Silver	6.17×10^7
Copper	5.80×10^7
Gold	4.10×10^7
Aluminum	3.82×10^7
Tungsten	1.82×10^7
Zinc	1.67×10^7
Brass	1.50×10^7
Nickel	1.45×10^7
Iron	1.03×10^7
Phosphor Bronze	1×10^7
Solder	7×10^6
Carbon Steel	6×10^6
German Silver	3×10^6
Constantan	2.26×10^6

TABLE D.1 Continued

MATERIAL (METALS)	$\sigma - (\Omega M)^{-1}$
Germanium	2.2×10^6
Stainless Steel	1.1×10^6
Nichrome	1×10^6
Graphite	7×10^4
Silicon	1.2×10^3
Ferrite (typical)	10^2
Water (sea)	5
Limestone	10^{-2}
Clay	5×10^{-3}
Water (fresh)	10^{-3}
Water (distilled)	10^{-4}
Soil (sandy)	10^{-5}
Granite	10^{-6}
Marble	10^{-8}
Bakelite	10^{-9}
Porcelain (dry process)	10^{-10}
Diamond	2×10^{-13}
Polystyrene	10^{-16}
Quartz	10^{-17}

TABLE D.2

MATERIAL	RELATIVE PERMITTIVITY - ε_R
Air	1.0005
Alcohol, ethyl	25
Aluminum oxide	8.8
Amber	2.7
Bakelite	4.74
Barium titanate	1200
Carbon dioxide	1.001
Ferrite (NiZn)	12.4
Germanium	16
Glass	4-7
Ice	4.2
Mica	5.4
Neoprene	6.6
Nylon	3.5
Paper	3
Plexiglas	3.45
Polyetheylene	2.26
Polystyrene	2.56
Porcelain (dry process)	6

PERMITTIVITY

The relative permittivities for a variety of common dielectric materials are given in Table D.2. These values are typical at room temperatures and for low frequencies. More detailed specifications can be found in a variety of handbooks.

TABLE D.2 Continued

MATERIAL	RELATIVE PERMITTIVITY-ε_R
Pyranol	4.4
Pyrex glass	4
Quartz (fused)	3.8
Rubber	2.5 – 3
Silica or SiO_2 (fused)	3.8
Silicon	11.8
Snow	3.3
Sodium chloride	5.9
Soil (dry)	2.8
Steatite	5.8
Styrofoam	1.03
Teflon	2.1
Titanium Dioxide	100
Water (distilled)	80
Wood (dry)	1.5 – 4

PERMEABILITY

The relative permeabilities for a variety of common magnetic materials are given in Table D.3. These values are typical at room temperatures and for low frequencies. More detailed specifications can be found in a variety of handbooks.

TABLE D.3

MATERIAL	RELATIVE PERMEABILITY - μ_R
Bismuth	.9999986
Paraffin	.99999942
Wood	.9999995
Silver	.99999981
Aluminum	1.00000065
Beryllium	1.00000079
Nickel chloride	1.00004
Manganese sulfate	1.0001
Nickel	50
Cast iron	60
Cobalt	60
Powdered iron	100
Machine steel	300
Ferrite (typical)	1000
Permalloy 45	2500
Transformer iron	3000
Silicon iron	3500
Iron (pure)	4000
Mumetal	20000
Sendust	30000
Supermalloy	100000

Author Biography

Dr. David Voltmer, Emeritus Professor of Electrical and Computer Engineering at Rose-Hulman Institute of Technology, received his formal education at Iowa State University (BS EE), University of Southern California (MS EE), and The Ohio State University (PhD EE). Informally, Dave has learned much from his many students while teaching at Pennsylvania State University and at Rose-Hulman. Dave (aka Dr. EMAG) has taught electromagnetic fields and waves, microwaves, and antennas for nearly four decades. Throughout his teaching career, Dave has focused on improving teaching methods and developing courses to keep pace with technological advancements. Spare moments not spent with his family are occupied by long distance bicycling and clawhammer style banjo. Dave is an ASEE Fellow and an IEEE Life Senior Member.

Printed in the United States
by Baker & Taylor Publisher Services